T0073335

Star Power

Star Power

ITER and the International Quest for Fusion Energy

Alain Bécoulet

translated by Erik Butler

foreword by Dennis G. Whyte

The MIT Press

Cambridge, Massachusetts | London, England

The MIT Press would like to thank the anonymous peer reviewers who provided comments on drafts of this book. The generous work of academic experts is essential for establishing the authority and quality of our publications. We acknowledge with gratitude the contributions of these otherwise uncredited readers.

This book was set in Stone Serif and Stone Sans by Jen Jackowitz. Printed and bound in the United States of America.

Library of Congress Cataloging-in-Publication Data

Names: Bécoulet, A., author. | Butler, Erik, 1971- translator. | Whyte, Dennis G., writer of foreword.
Title: Star power : ITER and the international quest for fusion energy / Alain Bécoulet ; translated by Erik Butler ; foreword by Dennis G. Whyte.
Other titles: Énergie de fusion. English
Description: Cambridge, Massachusetts ; London, England : The MIT Press, [2021] | Translation of: L'Énergie de fusion. | Includes bibliographical references and index.
Identifiers: LCCN 2021000495 | ISBN 9780262046268 (hardcover)
Subjects: LCSH: Fusion reactors. | Controlled fusion—International cooperation. | MESH: International Thermonuclear Experimental Reactor (Project)
Classification: LCC TK9204 .B4313 2021 | DDC 621.48/4—dc23
LC record available at https://lccn.loc.gov/2021000495

10 9 8 7 6 5 4 3 2 1

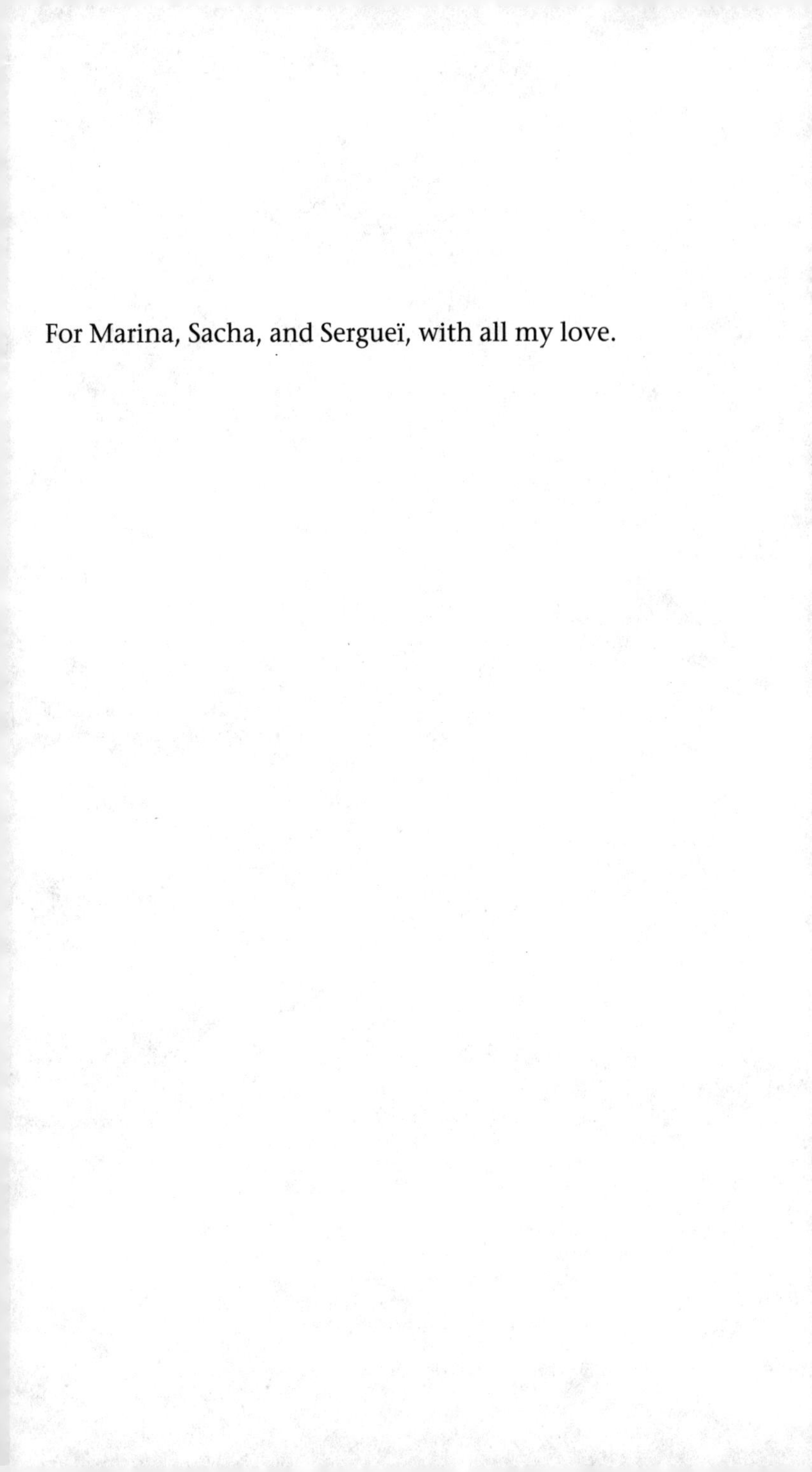

For Marina, Sacha, and Sergueï, with all my love.

Contents

Foreword

Dennis G. Whyte[1]

It is simultaneously fascinating and daunting how fast "things can change" in human society. As I write this, the world approaches one hundred million cases of COVID-19, an invisible virus that has brought staggering societal and economic changes in a single year across the globe. Yet several COVID-19 vaccines were developed and deployed in less than a year, a medical scientific feat that seemed itself unimaginable a year ago. In the span of the last few days the United States rejoined the Paris climate accord and the World Health Organization and the United Kingdom left the European Union. These are both positive and negative reminders of how rapidly humans and the societal structures we build, along with our collective worldview, can be adjusted.

1. Dennis Whyte is Director, Plasma Science and Fusion Center; Hitachi America Professor of Engineering; and Professor, Nuclear Science and Engineering, at MIT.

However, there are also fundamental truths unaltered by the near-term churn of events, two of which are the subject of this book, *Star Power*, by Dr. Alain Bécoulet. First, fusion is the fundamental energy source of the universe, making life possible everywhere. Second, the deployment of energy is at the heart of all human activity and achievement. Dr. Bécoulet has provided a compelling narrative, free of technical jargon, that describes these truths and the decades-long quest for realizing fusion energy on earth. Fusion energy systems could transform the world, so it is critical that a broad societal spectrum is informed about it. Yet fusion may seem like science fiction to some, intimidating to others, and generally based on an impenetrable science to all but experts, which does not bode well for an informed dialogue. Fusion scientists and developers must remove impediments to this dialogue by providing a respectful and accessible accounting of their activities. Alain Bécoulet has done admirable work in peeling away the mysteriousness of fusion science and technology.

Another underlying truth is that we are significantly altering the world's climate because of atmospheric carbon increases. This is mostly due to using fossil fuel energy sources, and therefore the truth about humanity's energy use provides a perplexing but inextricable link; the very energy source we are using to elevate and expand our capabilities can cause a fundamental change to our environment that will threaten the same. And while the exact details of this threat are being refined, we have consensus that we are running out of time to make the wide-scale energy changes required. The endeavor of decarbonizing our energy sources by midcentury is daunting. It is much more than decarbonizing our electricity sources, which accounts for less than a quarter of emissions. Other emission sources, including long-range transportation, agriculture, cement production, and industrial processing, are likely more difficult to decarbonize because they require more than just carbon-free energy sources. Certainly there are options on the table for decarbonizing, but they face challenges: renewables have advanced rapidly in cost-effectiveness and performance but have fundamental limitations of relatively low power density and intermittency, hydroelectric power has limited geographic deployment, and nuclear fission faces acceptance and economic impediments. There are growing efforts to tackle these challenges, but it is far from clear or certain they can be surmounted.

Fusion plays a special, if not unique, role in this mix. The science of fusion features on-demand, high power density, effectively limitless fuel, requiring no special access to a raw resource, with safety features that will permit wide acceptance—features that allow one to extrapolate it as a sustainable, indefinite carbon-emission-free power source for humankind. Conversely, it is the one energy source which has not been directly demonstrated as a viable commercial power source. Quite the combination! So how might one address this mismatch of need and opportunity? Looking at fusion's history, the oil crisis of the 1970s aligned with the advent of the tokamak's science success and saw the launch of resources and innovation that had fusion performance beating Moore's law for its pace. Yet the press of 2050 decarbonizing was not in play throughout almost the entire era of ITER. Almost a direct consequence is that an "ITER only" approach to fusion guarantees that fusion will not be in play as an energy source until the end of this century, as stated by Dr. Bécoulet. Given fusion's long-term prospects as an energy, this is a worthy pursuit. But it lacks the value proposition for fusion to play a meaningful role in humankind's largest energy transition. Unsurprisingly, the pursuit of fusion has evolved rapidly over the last decade to fill this vacuum, attempting to meet fusion energy's technical and economic requirements faster.

Like any complex integrated advance in knowledge, fusion has always progressed through bootstrapped and coupled science and technology breakthroughs. Dr. Bécoulet has outlined several of these, including the development of the tokamak concept in Europe and Russia. Others, mostly notably the United States and Japan, also provided key advances found in ITER. The JT-60 and DIII-D programs advanced new conceptualizations and technologies for the prospects of operating tokamaks in steady state. Locally, our own series of experiments at MIT, the Alcator projects, proved the viability of refractory metal walls and advanced options for exhausting the plasma heat.

The most important technology in magnetic fusion is the magnetic field produced by electromagnets. The science of confinement is such that as the magnetic field increases, which improves the insulation provided by the field, the fusion plasma can be made much smaller. Specifically, the fusion power per plasma volume, the key metric in economic attractiveness, increases as the magnetic field to the fourth power. Therefore, at fixed plasma physics conditions a doubling of the magnetic field allows the plasma volume to be sixteen times smaller, yet produce the same fusion power. The Alcator approach was a synergy of specialized high-field cryogenically cooled copper electromagnets with tokamak science, resulting in the first demonstration of the density and confinement time crossing the Lawson criterion. In 2016 the final tokamak in this line, Alcator C-Mod, set the world record for the product of the plasma density and temperature, surpassing two atmospheres of pressure for the first time. Alcator's very high magnetic field allowed these accomplishments in highly compact devices, factors of ten to a hundred smaller in volume than most of its peers, and almost a thousand times smaller than ITER. Yet this promising approach was limited for fusion energy because superconductor magnets could never survive the high magnetic fields required.

Into this scenario a breakthrough new kind of superconductor material called "high-temperature superconductors" (HTS) became commercially available in the

2010s. HTS is highly tolerant to high magnetic fields and so made feasible a new class of electromagnets which themselves would open new vistas for magnetic fusion. The connection between magnet technology and fusion is not new. ITER was made possible by the development of niobium-tin superconductor magnets, which approximately doubled the field of the preceding superconductor magnets made from niobium-titanium, familiar to most as the source of magnetic fields in MRI. A decades-long development program ca. 1990 took this nascent niobium-tin technology from conceptual designs all the way up to large-scale demonstrations. For example in the 1990s an international team involving the United States/MIT and Japan built a model central solenoid, the largest pulsed superconductor coil ever produced, and this magnet technology directly enabled ITER's tokamak design to be made relatively smaller because of its high magnetic field.

The opportunity afforded by HTS magnets was assessed by many groups but generally pointed a way to a dramatically smaller tokamak to meet the plasma confinement requirements. A prime example was ARC, designed chiefly by young scientists and students at MIT, which could produce ITER's five hundred megawatts of fusion power in a tokamak about the size of JET, that is, much smaller than ITER. A first pass at the cost of ARC showed that its normalized cost of power, namely, the power output per construction cost, was compelling as a "pilot plant," so that it was reasonable to consider ARC as the first entry point for fusion in the marketplace. This exercise then expanded to consider the development path for the HTS magnets with a private sector spinout: Commonwealth Fusion Systems was launched to attract investment resources, with the goal of rapid development of fusion energy on the grid. The MIT/CFS effort, backed by major energy companies like Eni and high-profile clean energy investors like Breakthrough Energy Ventures, has addressed the integrated design issues for the HTS magnets while simultaneously designing the SPARC tokamak. As documented in a series of peer-reviewed papers released in October 2020, SPARC projects producing one hundred forty megawatts of fusion power and energy gain greater than two (Q>2) using a magnetic field double that of ITER. Using the performance rules established for ITER, the most expected result is an energy gain of approximately eleven (Q~11). Yet SPARC has a volume about forty times smaller than ITER, a direct result of the fusion power relating to the magnetic field strength to the fourth power. So just as ITER's design was optimized with the previous generation of superconductor magnet advances in the 1990s, so again the tokamak fusion design optimization has changed because of HTS technology that was previously unavailable.

The apparently disruptive influence of this new technology, developed only in the last year or two, can change the timeline to fusion energy. Part of this is being able to have focused, smaller teams and backers to deliver the technology and projects. SPARC is planned to start operations in 2025, which is within the historic experience of building similar-sized fusion devices. Obviously, the prospects for near-term commercialization through a pilot plant are greatly improved by HTS. I personally have never been more excited and optimistic about the future of fusion energy.

This development is not surprising to those familiar with the uneven pace of development for new technologies. Twenty years ago, the mapping of the human genome was accelerated over tenfold by the synergistic combination of a higher risk approach with newly available computing. Yet the innovation there was not just in the technology but also in the organizational structures used to achieve a worthwhile and valuable goal. Moving this new approach into a private venture altered the risk tolerance and expected timelines compared to a public program. More recent examples include the success of the private sector SpaceX, decreasing substantially the timeline and cost for low-orbit launch, and the one-year development of several COVID-19 vaccines by multiple companies. Through our joint MIT/CFS effort of the last several years, I have seen up close the power of this focused approach. Yet such successes, including our own, are built on the foundations of public funding of basic research. The optimal organizational and resource deployment model for technology development is the subject of an ongoing debate.

Fusion energy's development appears to be at a pivotal point. The fusion private sector has rapidly grown to include more than a dozen companies backed by nearly two billion dollars in investment, featuring a wide array of risk retirement strategies to fusion. While the underlying fusion science requirements are set by nature, these companies are attempting a wide variety of approaches. Interestingly, many of these are pursuing pathways that were largely abandoned by the large public fusion programs, but pathways now renewed by technology breakthroughs, like HTS magnets or new insights gained by plasma simulations. Despite their very different approaches, these new companies have banded together in a Fusion Industry Association for mutual support to provide a unified voice to the public/government sector both to gather support for private-public R & D partnerships and to seek regulatory certainty for fusion from governments. This latter point is critical because the regulatory agency and strategy itself will vary across nations and technical approaches to fusion; while ITER has provided an important first example of fusion regulation, the overall development of the fusion regulatory framework has just begun. Simultaneously, the last year has seen a pivot in government/public fusion development programs, including major new efforts toward building a fusion pilot plant: CFETR in China, STEP in the United Kingdom, and the pilot plant in the United States, with the latter defined by recent reports from the US National Academies.[2] These efforts all seek to bring forth significant advances in fusion capability in parallel with, and on similar timelines to, ITER.

Simply put, the fusion development landscape has completely changed in just a few years, mostly driven by the previously discussed urgency around tackling climate change. The fusion landscape now features a variety of technical, organizational, and funding approaches, and as a fusion enthusiast, I view this as a very welcome development. Innovation in science, technology, and funding models will be required across the board. It is important to remember that there are many ways to get to fusion energy systems and that getting to net energy or the first commercial demonstration of fusion is hardly the end of innovation. Aviation and aeronautics hardly stopped developing new ideas and technologies following the Wright Flyer or the DC-3!

The nonchanging truths of energy use and fusion have more closely aligned with the brutal truths about climate change. This looks like a race, not the companies or governments racing against each other, but in the race against climate change, so that fusion energy

2. National Academies of Sciences, Engineering, and Medicine, *Bringing Fusion to the U.S. Grid* (Washington, DC: The National Academies Press, 2021), https://doi.org/10.17226/25991.

can play a meaningful role. Newly available scientific and technical tools are in front of us to make that achievable, and history will judge us poorly if we do not seize this opportunity.

Foreword to the French Edition

Is there anyone who hasn't dreamed of an inexhaustible, safe, and plentiful energy source without a hint of greenhouse gases? Has anyone dared to dream of recreating the sun down on earth and controlling the power that nature generates at every moment in the heart of the stars above?

Rising to meet such a challenge might seem impossible, yet it's essential for the future of humankind. Energy lies at the heart of life, and its abundance means well-being, increased life expectancy, mobility, communication, and better working conditions. Just imagine a single day without energy other than what our bodies provide. Turn off the lights, don't use transportation, and eat nothing that took fossil fuels or electricity to be produced. . . . It might sound impossible, but there are millions of human beings who endure conditions unacceptable to us every single day.

Nuclear fusion represents part of the solution to providing all of humankind with more energy—and energy that's environmentally friendly, at that. To obtain the amount of power that one nuclear reactor generates over the course of a year with twenty-five tons of uranium, it takes millions of tons of coal, oil, or natural gas. Compare that to a mere three hundred fifty kilograms of hydrogen isotopes!

Putting the sun in a box isn't easy. The book at hand takes us down a seemingly impossible path to possibilities at the final frontier of science and technology. With authority and passion, Alain Bécoulet shows what fuels a mad endeavor that thousands of researchers and engineers are now pursuing. Together, we learn all about ITER, the greatest project in the scientific world. This undertaking has brought together Europe, Japan, China, India, South Korea, Russia, and the United States—all joining forces to build a machine of universal progress. Every day, the men and women working here are giving their all to go beyond standing borders and do what no one has ever done before.

Setbacks and obstacles are inevitable, but we have no alternative. As Dante Alighieri famously declared, "Look where you have come from. You are not made to live like brutes, but to follow the way of Virtue and Knowledge."

Gabriele Fioni
Director for International Cooperation
Atomic and Alternative Energies Commission
Saclay, France

1
A (Very) Brief History of Energy

One can define the realm of life and the living in many ways. In terms of system, each and every organism, animal or vegetal, is an entity that evolves over time by modifying its structure through what it finds in its surroundings. Thus, a plant synthesizes sugars from the water and oligoelements it extracts from the soil, the carbon dioxide in the air, and, through photosynthesis, the light of the sun. Animals regulate energy in order to chase prey and sustain themselves by ensuring a supply of proteins, sugars, and lipids.

It's easy to see that such processes, as slow and imperceptible as they may be, require that a given organism use one or more mechanisms—biochemical ones, for the most part, but also others—to interact with its environment. Likewise, and also on an intuitive level, it's clear that life is based on consuming the amount of energy necessary and sufficient for making the mechanisms in question work. If no energy is available, no life can exist.

In more general terms, and beyond the sphere of living organisms, physics teaches us that energy consumption fuels all activity; this is what puts things into motion and produces all changes, whatever their nature. This same energy, as we will see, may also be considered a good to be produced, transported, traded, stored, and, finally, consumed—like water or food. In fact, it has the same strategic and vital importance as what we eat. In equal measure, it's necessary for the existence and development of both the individual and society: the latter may be viewed as a complex, living organism that produces and consumes energy in order to thrive. The connection between energy and consumption is endemic, then, and it represents a key factor for analyzing and understanding how our societies evolve. In a wide array of studies, researchers have demonstrated the correlation between growth and energy consumption. Thus, a recent economic study from the University of Utah examining the period between 2005 and 2016 has identified near-perfect proportionality: "When economic growth rate rises above zero, an increase in economic growth rate by one percentage point is associated with an increase in primary energy consumption by 0.96 percent."[1]

Ever since human beings first set about extracting energy from their surroundings, they have employed their imagination and powers of observation. When only the mechanical energy of the body was available, people sought to supplement it by enlisting energy from animals. The textbook example is yoking a bull to a plow. In this manner, people discovered they could augment their forces by expending a little bit—providing their own labor and fodder—in order to get more in return from the animal. A horse can perform ten times as much work as an adult man is able to provide. Such progress made it possible for laboring individuals to spend less time in the fields while securing the subsistence necessary for themselves and their families. By the same token, the productivity of cultivated land increased, which opened the prospect of trading surplus foodstuffs for other crops or services. On this basis, by satisfying individual needs, human communities evolved and became groups with shared interests—first families and then villages, with roles defined for each member of the collective.

Hereby, *labor* came to represent a physical quantity: the amount of energy it takes to perform a task. In this context, work is the energy a beast of burden must provide to move a load over a given distance. As a rule, the amount of work involved is expressed in joules (*J*). One joule is the energy it takes to lift one kilogram ten centimeters off the surface of the earth, for example.

Human beings also learned to employ the other sources of energy nature provides. To stick to mechanical operations, a rock can be dropped from a certain elevation in order to break objects and obtain smaller, more useful ones. Observing this phenomenon, people not only recognized that gravitational force represents a potential source of energy; they also arrived at a practical understanding of *power*, that is, the speed at which a given quantity of energy is consumed. Simple experimentation is just a matter of lifting a rock to a certain height and then letting it fall under its own weight. Hereby, the energy balance stands at zero: as much energy is expended when lifting the rock as is released when it falls. At the moment of impact, however, all the energy is "consumed" very rapidly, with effects that vary in keeping with the object that the rock strikes. By this means, one can see how to manage energy: build it up, store it, and then expend it at will. *Power* refers to the difference in energy over a given period of time. It is measured in watts (*W*); one watt represents the variation of energy by one joule over the course of one second.

Humankind has long known that mechanical energy can be profitably drawn from sources other than gravity or animals. All civilizations along coasts and waterways have employed wind power for sailing. For centuries, wind has also been used to mill grain. In the course of exploring these possibilities, human beings began to consume more energy than they could generate themselves—even though, in so doing, they were still using only a fraction of the natural resources available to them.

The art of fire represents another landmark in the history of human evolution. Here, a new form of energy comes into view: heat. Fire introduces a very different kind of mechanical energy, one that is high performance and, as we will see, closely tied to chemical reactions—that is, to manipulating the force holding atoms together inside molecules. While this energy source provides only heat, learning to store and channel heat is what ultimately led to the invention of the steam engine, the motor of the first industrial revolution. Effecting spectacular changes on the level of society as a whole, the industrial revolution also marked a seismic shift in the rate of individual energy consumption. By the end of the nineteenth century, the average European was already using, daily and just for him- or herself, an amount of energy equal to the muscular labor of several dozen individuals. This development—which counted as a sign of virtue in its day—signaled a decisive phase in the merging of energy consumption, progress, and comfort.

Frequently, calories are used instead of joules to quantify energy present in the form of heat. One calorie corresponds to the amount of energy it takes to raise the temperature of one gram of liquid water by one degree Celsius. It's interesting to note that the calorie (or, more precisely, the kilocalorie—that is, one thousand calories) is the unit employed for counting the exchange of energy within our bodies; a great part of the body's activity involves transforming food into heat in order to maintain temperature.

In hindsight, we know that humankind's massive use of combustion—whether wood, coal, petroleum, or gas—has made levels of carbon dioxide and other "greenhouse gases" explode, which has had a major impact on the thermal equilibrium of the planet as a whole. The greenhouse effect represents an environmental time bomb, and its consequences have been felt more and more over the last few decades; now, more than ever, defusing this bomb represents a matter of utmost urgency. In the process, people have come to recognize that their appetite for energy is making them tax natural resources to a dangerous degree. Once again, hindsight reveals—especially in light of the large-scale oil crises at the end of the twentieth century—just how blind we have been to burn resources that, in some instances, took millions of years to form. The intoxication of "easy" energy and progress has produced unwelcome consequences.

Nor does humanity's addiction to fossil fuels stop there. In the second half of the twentieth century, the race for energy—which is indispensable for growth across the globe—became the principal reason for conflicts between countries and regions. A look at the history of the last few decades makes it clear that in the Middle East, for example, securing control over hydrocarbon resources (and their transport to shipping ports) has consistently played a formative role in geopolitical alliances and tensions.

The nineteenth-century domestication of electricity contributed to the upsurge in energy consumption. The host of innovations and advances it brought include light bulbs, new heating technologies, electric motors, and, more recently, the electronics and computer systems that have become just as necessary for daily life as water or bread. Even if, technically speaking, electricity is not an energy source, it forms a vector with a quasi-magical power to travel across immense distances, almost at once. That's why its use has been increasing exponentially over the globe for the last century and a half. Wherever electricity is introduced, it opens new possibilities for development, prosperity, and communication—in a word, growth. Today, energy consumption is counted on an individual scale and measured in *kilowatt hours* (that is, how many thousands of watts are consumed in the course of an hour).

Electricity gave, and still gives, people excellent reason to increase their energy demand. Personal use already equals the muscular expenditure of several hundred individuals per day!

Even though appliances themselves don't produce greenhouse gases, this is hardly the case for the processes that generate, store, and transport the electrical energy they use. Today, the carbon footprint of one electric vehicle—its manufacture and use—can be greater than that of a gas vehicle in countries where electricity is generated in coal-fired power plants.

These introductory remarks have only touched on the daunting complexity posed by energy in our times. Energy is a matter of quantity: a collective good to be generated, stored, made available to the public, and ultimately consumed. Strategic questions are compounded by matters of grave—indeed, vital—importance. The three major issues concern long-term resource management, the future of the earth's climate, and geopolitical stability. It should be self-evident that access to energy, like access to water, represents a fundamental, universal right. At the same time, it's necessary to respect the environment and consider the immense commercial and political interests at play in the modern energy market.

With this in mind, we now turn to the great adventure, past and present, of research and development in fields connected to energy. As we have seen apropos of electricity and petroleum, development can precede demand. More and more, however, innovations are being made in order to find solutions in the spheres of production, transportation, storage, and/or consumption. Today, everything is headed in the same direction: toward global optimization, which takes all these factors into account in specific economic contexts. For years, intergovernmental agencies such as the International Energy Agency, the International Atomic Energy Agency, and Euratom (in Europe) have promoted cooperation between states. However, the law of supply and demand remains the principal motor of activity. For proof, one need only consider the political courage that must be shown—again and again—to incorporate environmental factors (even though they are more than evident to all and sundry) when working toward an ideal arrangement for producing energy. Many considerations are planetary in scope—for instance, the effects of greenhouse gases or the consequences of large-scale nuclear accidents. Others are generic in nature, such as insulating buildings. Others still are local and highly particular. The optimal "energy mix" is not a single equation, then, but a host of possible solutions that bear on each and every individual and his or her daily habits.

The terms of the energy equation are well-known. First comes the source, without which no solution is possible. It is important to understand, then, and from the outset, what *energy source* really means. Antoine Lavoisier, the father of modern chemistry, is credited with the maxim "Nothing is lost, nothing created, everything is transformed." His dictum applies to the matter at hand. In effect, all energy derives from transformation—or *interaction,* as the physicists put it. Thus, when coal is burned, the chemical reaction between carbon (C), the essential component of coal, and oxygen (O_2) in the air makes elements recombine into molecules of carbon dioxide (CO_2), the product of combustion. The reaction releases a certain amount of energy as heat, which means that the final molecule will be more stable than its individual constituents. In the case of mechanical energy, however, identifying the primary source sometimes proves more difficult. The source may be the energy of gravitation, employed directly or indirectly, transformed into kinetic energy, that is, motion. This is what happens when falling water turns the wheel of a mill; tidal power plants that harness the sea to move turbines offer another example. Winds and marine currents can also be employed to create kinetic energy. These forces result from temperature differences between day and night and between various parts of the globe (say, the equator and polar regions). In this instance, the primary energy source is not the wind or a current itself so much as the sun, which heats the earth's surface to varying degrees at different times. We will discuss the matter in detail below.

These simple examples demonstrate that there's always a primary source of energy over and above the others, and that it's tied to an actual, physical phenomenon. In turn, one or more vectors serve to transmit such energy and make it available for human use. The primary energy source may lie hidden (or even outside the atmosphere of the planet we live on). Take the internal combustion engine. Burning a hydrocarbon serves to make a motor rotate, which yields translational motion when transmitted to the wheels of a vehicle. In a further step, systems can be devised to recuperate the vehicle's kinetic energy in the course of deceleration, in order to improve overall efficiency. In this context, Lavoisier's thermodynamic principle represents a real touchstone: energy can only be extracted inasmuch as it has been "put there" in the first place. Energy is a quantity that starts from key sources and then, in the course of further transformations, turns mainly into heat. Strictly speaking, no energy is "renewable"; we should always remember to conserve it as much as possible.

Let's take a closer look, then, at processes that occur naturally and offer primary sources of energy. There's no need to get into the details of classical or quantum physics, much less the physics of relativity. It's common knowledge that so-called elementary particles, grains of matter and light, constitute the basic building blocks of everything that surrounds us. Each light particle, or photon, is characterized by zero mass and a certain frequency of oscillation, and it carries energy that is directly proportional to its oscillating frequency. Likewise, each elementary particle of nonzero mass—electrons, protons, and neutrons provide the most familiar examples—carries energy proportional to its mass, in keeping with Einstein's formula, $E = mc^2$ (wherein c is the speed of light in a vacuum).

Each atom of matter is composed of these elementary particles: a nucleus—an arrangement of protons and neutrons—surrounded by a cloud of electrons. So-called strong interaction ensures the cohesion of the nucleus, which represents the first source of fundamental energy. On another level, each molecule is an arrangement of atoms sharing part of an electron cloud, which "welds" them together. This *electromagnetic* force, which ensures their cohesion, is the second type of fundamental energy, which governs chemical reactions. In turn, the potential energy of gravity takes advantage of the gravitational interaction that exists between all objects possessing a mass. As surprising as it may sound, all the sources of primary energy we use originate in the basic bonds within nuclear or atomic structures; energy is extracted when we modify these structures by initiating chemical or nuclear reactions or by enlisting the force of gravity. This is also the case for so-called renewable energy sources.

We have already noted the "chemical" nature of combustion (wood, coal, gas, petroleum). We can now add forms of energy derived from reactions that rearrange the nucleus of atoms, that is, nuclear energy from fission harnessed in power plants, as well as solar energy (photovoltaic or thermal) that originates with thermonuclear fusion in the sun. Indeed, wind power belongs to the second category because it derives from solar energy, which heats masses of air and causes them to move over the globe. Geothermal energy originates in the core and mantle of our planet and therefore results, indirectly, from effects of gravitation. To get back to *renewable energy*, this term, which corresponds to the perspective of users, is anthropomorphic and egocentric inasmuch as it is tied to our impression that we are simply drawing on natural resources; in fact, it names a limited scale that we happen to be able to understand. The first principle of thermodynamics ("Nothing is lost . . .") is there to remind us that renewable energy does not exist in strictly scientific terms.

One piece of information about primary energy sources still needs to be added: the orders of magnitude for the processes at issue. The binding energy of an electron circling a nucleus is a million times smaller than the energy connecting protons and nucleons to each other inside the nucleus. Extracting energy from the core of nuclei is synonymous with efficiency, then: being able to make extremely compact and powerful sources of energy. Nuclear fusion research has taught us that the sun uses one gram of hydrogen to make as much energy as may be obtained by burning eight tons of petroleum. In this light, the appeal of the "game" that fusion researchers are trying to play is quite clear.

Apropos of the ideal energy equation—that is, considering the best-possible arrangements for production, transport, storage, and consumption—the task before us is to optimize each factor in order to obtain a situation that is not "renewable" so much as "durable." In effect, *durable* is the preferable term because it highlights the importance of economical and farsighted resource management.

In order to last, energy production should be *efficient*. In other words, as few resources should be consumed as can be added at any given level of operation, and this should occur in as clean a manner as possible. Transportation should proceed with minimal loss per unit of distance, consume few resources, and minimize environmental impact; in general, then, the distance traveled ought to be short. (Here, the difference from the limitations affecting primary sources is evident.) Storage ought to be quick, compact, easily repeated, and long-lasting; environmental considerations are important, too. Finally, consumption should be optimized so that each instance of use expends as little energy as possible, and at a fair cost to consumers. Coordinating all these factors requires a global approach in terms of efficiency, commercial viability, and social acceptability.

It's not easy to evaluate the relative merits of various sources of energy, promote one vector over another, or discount the role of research and development (that is, the factor of time). Doing so, one runs the risk of errors on a global scale. Batteries for electric cars represent a case in point. At present, their manufacture produces as much greenhouse gas as—if not more than—their use in vehicles saves over time, and that doesn't include the massive amount of rare-earth metals that this technology requires. For the time being, such batteries aren't very durable or appealing in geopolitical terms. That said, it would be a mistake to conclude that we should give up on electric vehicles; major advances have been made in the last few years to resolve these sticking points. Conversely, if fossil fuels have become considerably more attractive over the last decade, the development has occurred thanks to research on petroleum and gas, as well as improved techniques for extracting and purifying fuels. Over the last few decades, research has really taken off and advances are evident at each link in the energy chain, from the level of producers to that of consumers.

Thermal power plants and combustion motors have charted major progress, both in terms of energy efficiency and cleanliness. Coal power stations, which still generate about a third of primary energy and 40 percent of electrical energy worldwide, are known to produce the greatest amount of CO_2 by far. Still, technological advances and so-called clean coal processes (which increase the efficiency of combustion) have reduced environmental impact significantly. More and more, the coal industry is counting on trapping CO_2, and research in this area is being actively promoted. But even though improved technologies are being implemented in new plants, many more facilities exist that were built on the old model; the cost of modernizing them increases the price of energy. Natural gas plants—the most flexible and (in terms of investment) the cheapest way to supply electrical networks—have made spectacular progress both in cleanliness and yield thanks to so-called combined cycle technology, which connects a steam turbine to traditional gas equipment in order to use the heat generated as fully as possible. Needless to say, the gas industry is not exempt from the questions surrounding CO_2 production; it will benefit from further advances—better storage technology, for example.

Where the internal combustion engine is concerned, the dramatic decrease in the quantity of petroleum consumed by automobiles shows that heat is being transformed into mechanical propulsion much more efficiently now. Lower rates of consumption obviously mean less greenhouse gas and fewer microparticles that threaten health. However, recent events have also shown that manufacturers occasionally have to adjust engines to meet environmental standards that are not necessarily in phase with what is technologically or economically viable. It would be wrong to go on a witch hunt. Supply and demand are twin sisters.

The chapters to follow will discuss nuclear power and the operations of nuclear facilities at length. Here, too, research over the last few decades has yielded considerable advances in terms of productivity and safety. Energy conversion, especially between alternators and the electricity that comes out in the end, has also improved markedly. Thus, equipment and rotation speeds are now better matched to specific sources, with very slow conversion rates for hydraulic, tidal, and wind-powered plants.

Likewise, converting the energy of solar rays to produce electricity by photovoltaic means has taken off thanks to aggressive and competitive research. Relevant programs involve increasing output (which still remains low) and improving the design of photovoltaic panels. These technologies are being integrated more and more into daily life, using different surfaces and making panels transparent or flexible. New materials other than silicon enable a conversion yield above 25 percent. Sites of installation include roadways, roof tiles, and everyday objects with built-in photovoltaic features. There are even plans for production units that concentrate light or follow the sun's path over the course of the day in order to maximize efficiency. The Swiss experiment of Solar Impulse—two planes that have traveled around the world—has demonstrated that airborne propulsion also lies within reach. Finally, and to be sure, solar ovens make it plain that the sun's energy can be captured directly and turned into heat without photovoltaic conversion. Multimegawatt plants are already in operation in Mediterranean countries like Spain, and large-scale projects are in the course of development in Morocco, among other places. *Solar power* has become one of the cheapest sources of energy across the globe.

Improvement to energy vectors is proceeding more slowly and in a less spectacular manner, but the signs are encouraging. In effect, vast networks of wires for distributing electricity are still in place. The energy loss that occurs in this process depends on the electrical current being transmitted; in order to minimize loss, energy must be transported at a high voltage, which means ensuring a certain distance between cables and the ground. One of the most interesting fields of research now concerns so-called smart grids, which seek to integrate production, transport, and consumption as closely as possible. To this end, it's necessary not only to develop infrastructure—sources, networks, storage facilities, proper insulation of buildings, and so on—but also to analyze material conditions and calculate likely rates of consumption. Doing so opens the prospect of whole cities that will be more or less autonomous in terms of energy throughout the year. Tapping into the national grid would only occur in an emergency.

As for energy vectors intended to replace petroleum products and their transport, we should note the importance of hydrogen (H_2), which offers the singular advantage of not producing greenhouse gases when it burns with the oxygen present in air. This vector holds great promise for the vehicles of tomorrow and currently is the object of intensive research across the globe. Making and distributing hydrogen still poses a major difficulty, especially given the high level of primary energy sources necessary to synthesize it. All the same, on the national and regional level, hydrogen production will find a place in all prudent economic planning; over time, any surplus will be "consumed" entirely or in part.

Innovative measures are being taken to use energy more economically. Efforts charted a certain success in France during the two gas crises at the end of the twentieth century. Readers might still recognize the famous slogan "In France, we might not have gas, but we do have ideas" (*En France, on n'a pas de pétrole mais on a des idées*). In effect, some estimates hold that about 50 percent of energy gets "lost" through inefficient fuel systems (vehicles, buildings, and so on). Housing provides a striking example. It is now possible to construct buildings that consume practically no energy—for instance, by capturing sunlight over the course of the day or by means of high-performance insulation that maintains living quarters at a constant temperature without additional heating (or, on the coldest days of the year, with just a few radiators or space heaters). The best way to reduce energy consumption per capita is, of course, to avoid waste at the individual level.

In spite of great progress and worldwide research, energy production still poses daunting challenges that must be resolved both for the stability of individual societies and for the overall welfare of our planet. Economic and geopolitical considerations dominate the market as never before. In addition, demand is only growing, and it is certain to increase even more in coming decades for two major reasons: the expanding world population and the desire in developing nations to reach standards of living equivalent to what Western societies enjoy. These factors mean that the search for primary energy sources will continue to mount in intensity, as will concerns about environmental limitations, natural resources, and international balances of power.

It is time, then, to shift our discussion to nuclear energy, which a priori offers great advantages in these three areas. From the outset, we should situate nuclear energy in a global context and call to mind, by way of a few examples, the reasons why it is vital for human societies to have massive amounts of energy available, over and above sources that can be readily installed near sites of consumption (photovoltaic panels, wind power, or small, gas-powered generators, for instance). Industrial activity is the first reason. The immense amount of electricity used by plants that extract and process raw material, or in the process of transporting goods by rail, makes it necessary to distribute hundreds of electrical megawatts over significant expanses (for example, from one end of France to the other). Large urban areas consume a great deal of energy (for lighting, heating/climate control, and so on), but they do not have enough space available for setting up photovoltaic panels or windmills on an appropriate scale. Conversely, it's difficult to imagine power plants much larger than those currently in existence (ranging in output from a few hundred megawatts to several gigawatts), since transporting the electricity they produce grows more and more complicated as power and "wire loss" increase—to say nothing of the fact that the public dislikes high-voltage lines near homes.

To address the production of electrical energy by means of nuclear reaction, we'll start with how nuclear fission operates. The cohesive energy of nuclei—which is what enables their components (protons and neutrons, generically called *nucleons*) to form a structure—depends on the total number of these components. For nuclei heavier than those of an iron nucleus, cohesive energy decreases with the number of components in the nucleus. When these nuclei collide with other particles, such as neutrons, they break apart into smaller nuclei, each of which exhibits greater stability. The effect of this process is to set energy free, which is superior to what can be gained by means of chemical reaction (such as combustion); accordingly, it provides the basis for extremely high-yield sources of energy. The underlying principle of fission reactors is to recuperate the energy that reactions set free as heat, which generates steam used to move a turbine attached to an alternator; hereby, the primary source of energy, which is nuclear, is transformed into electricity, which can be more easily transported and used. At the moment, nuclear facilities employ uranium ore, which is present in the earth but not distributed uniformly over the globe; in other words, this natural resource is not exempt from geopolitical tensions or financial speculation. Yet uranium exists in abundance, which guarantees a bright future for the industry. Constant improvements to reaction efficiency and technologies for reusing fissile material ensure its long-term availability, with estimates now extending to millions of years. As far as the environment is concerned, nuclear fission does not generate greenhouse gases, which represents a further point in its favor.

Yet fission is responsible for a significant amount of by-products; most of them are radioactive and have a half-life[2] that can last for thousands of years—or even tens, or hundreds, of thousands of years. This fact must be taken into account, and it's important to understand every aspect of the so-called fuel cycle in order to weigh the societal and economic costs of the industry. Waste and storage also represent factors to be considered, both in economic terms and in light of the timescales involved.

Finally, on a geopolitical level, harnessing fission to civilian aims cannot be separated from eventually employing it to military ends. Even if the same principle applies to energy obtained by other means, it has a particular resonance where nuclear power is concerned. Fission yields a million times more energy than the chemical reaction of combustion, even if it's TNT (which is why the strength of atomic bombs is counted in thousands of tons, or *kilotons*, of TNT). Although monitoring organizations such as the International Atomic Energy Agency exist, one head of state or another, appealing to national sovereignty for access to nuclear energy, will regularly provoke a diplomatic crisis that makes headlines. Economically speaking, electricity from nuclear fission is very competitive since expenses are essentially a matter of infrastructure (investment and maintenance), not fuel. All the same, costs tend to rise over time because of the tougher security and safety measures that plants require.

If fission—like all sources of energy—relies on research in order to enhance its appeal and acceptability, another kind of reaction provides, a priori, a much better response to the inconveniences that attend nuclear power. In effect, if it's possible to produce energy by breaking apart large nuclei, one can also generate energy by fusing small nuclei. As we have noted, cohesive energy decreases relatively slowly for nucleons in heavy nuclei; in contrast, it increases—and rapidly, at that—in light ones. If one manages to domesticate and exploit the nuclear energy set free when small atomic nuclei fuse (especially the smallest of them, hydrogen and its *isotopes*,[3] deuterium and tritium), then the problem of radioactive waste with a lengthy half-life can be remedied—as can safety concerns at nuclear plants, the issue of the availability of resources, and geopolitical tensions surrounding fuel supply. This primary source of energy is what powers the stars in the sky. The rest of this book will discuss this other way of extracting energy from nuclei—a tremendous adventure of research and development that has been unfolding for half a century now, with an unparalleled level of global cooperation and to peaceful ends.

2
The Saga of Nuclear Energy

In a sense, nuclear energy got off to a bad start. The remarkable revolution in science and technology that started toward the end of the nineteenth century and expanded in the early-twentieth century assumed terrifying dimensions in the destructive rage of the First, and then the Second, World War. The atomic bomb has come to symbolize divine punishment for human beings and their thirst for knowledge, which has given them the means to kill themselves off once and for all. This modern myth of Prometheus still shapes the way people think. Anything one says about nuclear energy requires that one distinguish between military and civilian applications. In contrast, chemistry need not account for the assorted explosives that have been used in war and terrorist acts for hundreds of years now, and no one expects biology or medicine to provide a justification that bacteriological weapons exist.

The tumultuous debut of nuclear energy can be precisely dated: August 6, 1945, eight o'clock, sixteen minutes, two seconds, Japanese time, when the bomb christened "Little Boy" exploded directly above Shima Hospital in Hiroshima. In a fraction of a second, Little Boy unleashed the equivalent of fifteen thousand tons of TNT, instantly killing tens of thousands of civilians, leveling everything within a two-kilometer radius, and causing, for years to come, incalculable damage and illness from the blast itself, radiation, and, to a lesser extent, the ensuing contamination. In the most brutal way, the world at war discovered the unbelievable might locked away in the core of the atom. Even the bomb's inventors were surprised—and, in some cases, traumatized for life. Leó Szilárd, the Hungarian physicist who first conceived of the nuclear chain reaction and advocated its military use by the Americans, would later declare: "Suppose Germany had developed two bombs before we had any bombs. [. . .] Can anyone doubt that we would then have defined the dropping of atomic bombs on cities as a war crime, and that we would have sentenced the Germans who were guilty of this crime to death at Nuremberg and hanged them?"[1] This nuclear explosion, then, was the culmination of a mad race between the United States and Germany, which lasted for more than three years, to master and wield extraordinary power—a new clash of the Titans, or battle between Good and Evil.

To gauge the "madness" of this race, we should note that the very first test of a plutonium bomb took place in the New Mexico desert on July 16, 1945, a mere twenty-one days before Hiroshima. Just five days later, on July 21, President Truman officially gave the green light for the operation. Detonating Little Boy over Japan was the result of secret research the U.S. military had conducted from 1939 on, dubbed the "Manhattan Project"; in 1942, the government granted it almost unlimited support. Indeed, at the start of the Second World War, Leó Szilárd and Eugene Wigner had informed President Franklin D. Roosevelt that a new understanding of the uranium nucleus made it possible to develop weapons infinitely more powerful than conventional ones—and, moreover, that Nazi Germany was actively pursuing such arms. The side that managed to make them first would be in a position to crush its adversary. This is precisely what happened in August 1945, except that the bomb fell on Japan; Germany had capitulated three months earlier, without the Allies detonating an atomic weapon over Berlin.

Instead of entering into details about major and minor developments in the Second World War or the Manhattan Project, we should retrace the important stages of research and science that made the new weapon possible. From the time of ancient Greek civilization up to the end of the nineteenth century, physicists viewed matter as being constituted by elementary particles, atoms. The word's etymology (*a-tom,* "uncuttable") points to the idea of wholeness. Only as the twentieth century approached did the discovery of radioactivity cast doubt on this certainty. Matter, it turned out, is capable of spontaneously emitting particles and, in this manner, of changing itself. The new generation of physicists included Henri Becquerel, Pierre and Marie Curie, and Ernest Rutherford—to mention only a few pioneers. It was not until the period between the two World Wars that researchers concluded that the atom holds a tiny nucleus with a positive charge, surrounded by a chain of negative electrons, so that the whole forms a neutral structure.

This nucleus, research revealed, is constituted by two (and only two) types of elementary particles, or nucleons: the proton, which contains a positive electrical charge, and the neutron, which, as its name indicates, bears no charge. The simplest nucleus is that of the hydrogen atom, which comprises a single proton. Then come the nuclei on Mendeleev's periodic table: the helium nucleus is made up of two protons and two neutrons, lithium of three protons and three or four neutrons, and so on. The helium nucleus soon received the title of *alpha particle,* since it occurs in phenomena of natural radiation (for example, what Marie Curie observed in samples of uranium ore). In the case of lithium, the third element on the periodic table, there are two isotopes (that is, two nuclei with the same number of protons but a different number of neutrons), Lithium-6 and Lithium-7, both of which are stable. In fact, almost all the atoms on the periodic table can occur in the form of one or more isotopes displaying varying levels of stability.

The force that binds nucleons, positive or neutral, to each other is very different from the force that makes electrons orbit the nucleus. It is called *strong interaction*, in reference to the high coherence exhibited by stable nuclei and their ability to overcome the repulsive force that protons exert on each other. The cohesion of nuclei, scientists determined, depends on the balance between the number of protons and the number of neutrons. This observation led to the coining of the phrase "valley of stability." For small nuclei, the valley of stability corresponds to a similar number of protons and neutrons; for larger nuclei, it involves a much higher number of neutrons than protons. Stable uranium, for instance, contains ninety-two protons and one hundred forty-six neutrons. If a nucleus possesses a number of neutrons and protons too far away from the valley of stability, it will disintegrate by breaking up into pieces; in the process, it frees up energy in the form of particles that are ejected at great speed. This is the phenomenon of *natural radioactivity*. Experimentation has confirmed three major types of radiation, each of which corresponds to the emission of specific nuclear particles: *alpha radiation*, whereby the nucleus breaks up into one or more pieces while emitting a helium nucleus; *beta radiation*, where one or more electrons are ejected from the nuclear structure as it undergoes modification; and *gamma radiation*, where one or more photons are released.

The real revolution, for our purposes here, dates to a 1934 experiment performed by Irène and Frédéric Joliot-Curie in France. By bombarding stable atoms with alpha particles derived from natural radiation, they induced a new kind of radioactivity that continued even when the initial alpha radiation had ceased. The nuclei "irradiated" in this manner transformed into new nuclei, which were radioactive themselves. *Induced radioactivity* had been achieved.

Word of induced radioactivity spread very rapidly, and the phenomenon was widely documented by other researchers. Just a few years later—on the eve of the Second World War—scientists recognized that it was possible to generate reactions of induced radioactivity in a "chain." If the right material is chosen, when irradiated by neutrons it will disintegrate and emit neutrons of its own. When such a reaction brings forth more neutrons than it consumes, the pieces are in place for a nuclear reaction that will sustain itself—or even "run away"—provided that a sufficient amount of fissile material (so-called critical mass) is available and that more than one neutron is generated by each reaction in the series. As we have seen, it would only take a few years before the Manhattan Project reached its goal. Many researchers contributed to advances in the field, among others the German physicist Otto Hahn, Niels Bohr from Denmark, and Lise Meitner, who, because she was Jewish, had to leave Austria for Sweden in 1939. Of course, scientists were also hard at work in the United States; their ranks included Enrico Fermi and the aforementioned Leó Szilárd and Eugene Wigner. French researchers including Frédéric and Irène Joliot-Curie played a major role, as well.

The rush to domesticate fission in order to produce electricity started in the United States, the Soviet Union, and France immediately after the Second World War. The Manhattan Project had already represented an initial effort, to a certain extent. From 1942 on, work continued at the University of Chicago, where Leó Szilárd and Enrico Fermi assembled fissile products for the first time: uranium, in the form of metal and oxides, placed in layers on a "neutron moderator" composed of graphite. In this arrangement, which soon came to be known as an *atomic pile*, the moderator serves to slow—indeed, to capture—a portion of the neutrons produced by the nuclear reaction in order to prevent it from running away in uncontrolled fashion, as occurs in a bomb. This design prevents neutrons produced in the reaction from inducing further reactions and therefore promotes a steady process.

The fundamental principle at work in the very first reactors was to use naturally occurring uranium extracted from ore; in addition to the most plentiful isotope, with two hundred thirty-eight nucleons, it contains a small portion of uranium-235, which has three fewer neutrons per nucleus. When uranium-235 absorbs a neutron, it momentarily transforms into uranium-236. Uranium-236 is unstable and undergoes fission in different ways. Hereby, it can generate a nucleus of krypton-93 and a nucleus of barium-140, for instance, or a nucleus of strontium-94 and a nucleus of xenon-140. In the first case, three neutrons are set free, and in the second two. Needless to say, these elements open the possibility of chain reactions. (That said, uranium-235 is present at a level below 1 percent in natural ore. To be used for a chain reaction, the ore needs to be enriched to a level between 3 percent and 5 percent.)

France had no intention of standing at the sidelines. Recognizing the country's strategic interest in nuclear power for both military and civilian applications, General Charles de Gaulle had the foresight to create the Atomic Energy Commission (*Commissariat à l'énergie atomique*; CEA) in 1945. The first atomic pile in France, baptized "Zoé," was launched on December 15, 1948, at the Fort de Châtillon, near Paris. Frédéric Joliot-Curie, now the first commissioner of atomic energy in history, supervised operations, and President Vincent Auriol was their sponsor. Zoé used *heavy water*—water in which hydrogen atoms have been replaced by one of hydrogen's isotopes, deuterium—as a moderator.

At the same time, hundreds of laboratories and commercial enterprises started to look for the best way to design fission reactors in order to achieve an optimal combination of performance, power, safety, and reliability. The giants of the worldwide nuclear industry now came into being. Above all the rest towered Westinghouse. This American firm gained the upper hand in the market with a license at the origin of most plants built not just in the United States but also in France and China. In equal measure, efforts were launched across the globe to develop methods of enrichment. As a rule, such technology relied on gaseous diffusion (or gas centrifuges), which offered economic advantages, in particular. Other procedures based on the chemistry of uranium or laser beams also were objects of research.

Today, the world is home to many nuclear reactor networks, which vary in keeping with the kinds of fuel, moderator, and coolant employed. They fall into two major families, according to how the speed of neutrons controls chain reactions. Thermal, or moderated, reactors follow the principle of slowing the neutrons released but not absorbing them, which promotes the fission of uranium-235 or plutonium-239. Fast reactors do not slow neutrons; instead, they make heavy atoms present in the fuel (such as uranium-238 or thorium-232) undergo fission, which neutrons of lower energy cannot break down. In the latter case, scientists speak of "fertile material." Using uranium-238 directly offers the further advantage of not requiring fuel to be enriched beforehand. Fast reactors, which are also known as *breeder reactors,* generate fissile material from fertile material in excess of the amounts naturally present at the start of the cycle. As such, they open the prospect of much greater durability: the resources required will be available for thousands of years. Likewise, they have the potential to "burn" radioactive material generated by other kinds of nuclear reaction and can therefore "process" by-products (for instance, plutonium-239).

Some four hundred fifty nuclear reactors are in operation across the globe at the moment. Almost all of them are thermal reactors. For the most part, fast-neutron reactors are prototypes in the service of research and development.

For some time now, it has been standard practice to classify fission reactors by *generation*. Doing so enables us to distinguish between different stages of technological evolution and levels of safety. The first generation of reactors extends from the years following the Second World War up to about the 1970s. Examples include Caller Hall/Sellafield, a facility in use from 1956 to 2003 in England, and Chooz A, built in France by Framatome under license from Westinghouse, which was in operation from 1967 to 1991. The model of reactor known as "UNGG" (*Uranium Naturel Graphite Gaz*)—or, alternatively, "graphite-gas"—also belongs to the first generation. UNGG reactors were the first examples of French design; inspired by Zoé (the atomic pile), they were built at Marcoule, Chinon, Saint-Laurent-des-Eaux, and Bugey; operations ceased between 1968 and 1994.

Reactors built between 1970 and the end of the twentieth century represent the second generation. In France, the nuclear industry flourished during the gas crises; even now, plants dating to this period form the majority of reactors in operation. The same holds for most nuclear power stations across the globe. Second-generation facilities include pressurized water reactors for the most part, but also boiling water reactors and so-called advanced gas-cooled reactors.

The Chernobyl disaster in 1986 occurred on a graphite-moderated boiling water reactor of Russian design, the RBMK-1000 (реактор большой мощности канальный, or "high-power channel-type reactor"). The event revealed flaws associated with second-generation technology, as well as safety problems attendant on the use of equipment: missing provisions for confining radioactive material in the case of accident, manual aspects of operation that posed security risks, inadequate oversight, and, to be sure, suboptimal management of crisis situations. These shortcomings concerned not just the Chernobyl plant in particular but the nuclear industry as a whole—which entered the twenty-first century with a shaky bill of health. The more recent accident at Fukushima has revealed further risks associated with second-generation reactors, and at multiple levels of operation.

Stronger safety regulations, heightened monitoring of production levels, and improved international communications are in place for third-generation facilities. In addition, technical solutions have been developed to limit—if not eliminate—potential causes of accidents. Reactors designed from the 1990s on, in the wake of the Chernobyl disaster, have been scheduled to go into operation since the early 2010s. They include, in particular, so-called EPR reactors, which now are being built in Finland, France, and the United Kingdom; the first plant of this kind has just started up in China. The chief goals are increased safety and a higher rate of economic return.

In concluding this rapid—and necessarily incomplete—overview, we should note that the research community is developing a fourth generation of designs. For the most part, they are fast-neutron reactors, based on a conception and mode of operation rather different from their predecessors (although, obviously, they will benefit from safety improvements made in the third generation). The goal is to create reactors that will consume fertile materials and, in so doing, reduce the amount of waste that has been generated until now. The undertaking is not entirely new. A fast-neutron reactor has been up and running in Russia since the 1980s. In France, a prototype called *Phénix* has existed for more than thirty-five years; its industrial extension, *Super Phénix*, began operation in the 1980s, too, although it was closed down at the end of the 1990s for political reasons.

In 2011, the International Atomic Energy Agency in Vienna launched the Generation IV International Forum to promote and coordinate work in the field. In this context, some half-dozen projects have emerged, bringing together more or less all researchers working on fast reactors from across the globe. France is in charge of one of one of these initiatives, which has been baptized "ASTRID" (Advanced Sodium Technological Reactor for Industrial Demonstration). ASTRID will take up, and improve, key aspects of the Phénix reactor, in particular, the use of melted sodium for cooling.

At this juncture, it's worth pausing for a moment to discuss nuclear energy's public acceptability—or lack thereof—over the first seventy years of its history. Fraught with paradox, questions haunt plans even now.

Although the aftershocks of the United States' use of two nuclear weapons against Japan were felt for some time, the period following the Second World War was synonymous with Allied victory and a strong initiative to rebuild. From here on, conquest of the atom would symbolize power and growth—which, of course, also meant the struggle for supremacy. Only the five countries that possessed atomic technology obtained permanent seats on the extremely exclusive United Nations Security Council: the United States, the Union of Soviet Socialist Republics, the United Kingdom, France, and China. This board promptly barred access to nuclear weapons not just for defeated Japan and Germany but de facto for everyone else, too. Efforts to prevent proliferation had begun. Although it proved difficult to check the exchange of knowledge in an academic milieu, information was rigorously classified to stop technical know-how about weaponry from trading hands; likewise, by civil and military means (including covert measures), access to fertile and fissile material was blocked, as were efforts to enrich uranium-235.

Needless to say, the same period witnessed mounting rivalry between the major winners of the Second World War. The United States and the Soviet Union quickly began to steer a political and economic course in line with the new arms race. The number of weapons constructed on both sides soon warranted the title of "escalation," with "nuclear umbrellas" extending over the territories of the superpowers' respective allies. The "Cold War" of deterrence between two giants, waged by means of both truths and falsehood, lasted until 1991, when the Soviet bloc finally broke apart. Up to this point, any number of incidents occurred—sometimes with the prospect of direct nuclear deployment, as in the Cuban Missile Crisis (1962). The first fifteen years of nuclear energy, then, centered on the power of weaponry, and the international public had little chance to become aware of peaceful uses for the atom (or, for that matter, the drawbacks that might be involved).

Blocking access to nuclear energy barely lasted for a decade. At its 1958 conference in Geneva, the International Atomic Energy Agency, founded by the United Nations, launched a worldwide program called Atoms for Peace. Proper monitoring would provide a more realistic plan in the world now being rebuilt than wholesale bans. A large portion of nuclear research was declassified, giving rise to international cooperation that was robust, structured, and overseen by the International Atomic Energy Agency itself. Information made available concerned areas and technologies that would ensure peaceful uses of nuclear power; at the same time, strict measures of confidentiality were intended to prevent unauthorized parties from developing atomic weaponry. For the first time, the general public was made aware of the difference between military and civilian applications.

This brings us to the youngest but brightest member in the "atomic family": thermonuclear fusion, the same process that makes the stars shine. Whereas the first atomic bombs—generically known as "A-bombs"—relied on fission of uranium or plutonium, the escalation that ensued soon prompted governments and militaries to seek a weapon even more powerful. The new kind of bomb relied on fusing two hydrogen isotopes, deuterium and hydrogen, to produce significantly more energy than fission (which uses elements with large nuclei). The process here is not a chain reaction. Fusion requires the collision, at immense speed, of a deuterium nucleus and a tritium nucleus, which yields a highly charged nucleus of helium-4. Nothing in this reaction can give rise to further amplification. What's more, it can only occur when nuclei are traveling at high speed and collide frequently. The combination of deuterium and tritium must be brought to a state of intensive thermal agitation and prove sufficiently dense—that is, reach an extremely high pressure level (pressure being the product of the density of the medium and its temperature). The temperatures in question equal those found at the core of stars, ranging from tens to hundreds of millions of degrees. It's quite the challenge. The quickest and most direct way to obtain such temperatures and pressures is to place the mixture of deuterium and tritium at the center of an A-bomb;

the bomb's explosion will compress the mixture, yielding the conditions necessary to set exothermic fusion into motion. This is the principle behind the H-bomb, which the United States set about developing—in order to stay one step ahead of the competition—after the Russians detonated their first atomic bomb on August 29, 1949. The first American hydrogen bomb exploded on November 1, 1952. Before long, its Soviet counterpart followed.

In 1958, then, when the conference Atoms for Peace was held, the public did not yet recognize the difference between atomic and hydrogen bombs. But researchers were already working to use nuclear fission to peaceful ends and to harness the energy produced by nuclear fusion. When information pertaining to these two branches of nuclear physics was declassified, the gates opened for one of the greatest adventures of the human spirit and mind: controlled thermonuclear fusion.

3
Power from Fusion

Paradoxically, the principles of deriving energy from fusion were discovered before scientists knew what to make of radioactivity—that is, before nuclear fission was understood. At the same time, human beings have possessed technological control of fission for three-quarters of a century now, but they have not managed to do the same for energy from fusion. The matter merits discussion.

The essential principle of fusion is based on the capacity of nucleons (protons and neutrons) to assemble and form structures (atomic nuclei) in spite of electrical repulsion between protons, which all have a positive charge—the so-called Coulomb barrier. This circumstance is all the more curious given the fact that the Coulomb barrier increases as space between particles decreases; in other words, repulsive force changes in inverse proportion to the square of the distance between them. The *nuclear force* that enables nucleons to approach each other at close range, then, is based on another kind of interaction, which must be much "stronger" than electrical repulsion and operate "closer at hand." In effect, measurement reveals this force's intensity to be hundreds of times stronger than electromagnetic interaction between charged particles; its scope is very slight and, by nature, on the scale of the atomic nucleus (that is, one femtometer, or fermi—10^{-15} meters). When scientists first discovered this new interaction, they deemed it "fundamental." Subsequently, over the last quarter of the twentieth century, they determined that nucleons are not, in fact, elementary particles but rather made up of even smaller "grains": quarks and gluons.

In actuality, the force ensuring cohesion within atomic nuclei is the residue of the truly fundamental interaction that governs relations between quarks and gluons. At the most primal level of matter, each quark bears a symbolic "color charge" (three of which are possible: blue, red, and green) that can be exchanged with another quark by way of a gluon, which possesses a color and an anticolor of its own. Accordingly, a "blue anti-red" gluon will be able to change a red quark into a blue one when they interact. In this ultramicroscopic world, a given nucleon necessarily comprises three quarks, each of which bears a different color. At a sufficient remove, it will appear "white"—that is, "neutral" in the sense of strong interaction. The distance in question corresponds precisely to the scope of strong interaction intuitively introduced above. Strong interaction, then, is responsible for the cohesion of atomic nuclei on the basis of the forces it exercises between the constituents of nucleons when the latter are in reach of one another. If the distance between two nucleons grows too large, they may fail to respond to each other and separate.

Although the comparison is imperfect, an analogy to electromagnetic interaction will help us understand the concept of color charges—the nature of gluons, in other words. Particles sensitive to electromagnetism carry an electrical charge (electrical charges are only two in kind: positive or negative), and they interact by exchanging a bead of light, or photon, which is neutral (and can be understood as the copresence of both a positive and a negative charge). Particles that are not electrically charged, such as neutrons, are not susceptible to electromagnetic interaction.

But to understand fusion between nucleons in full, we need to add a third fundamental force to those at work at the atom's core: so-called weak interaction, which occurs at a level of intensity one million times lower than that of strong interaction, and with a scope a hundred times smaller. Even though weak interaction operates only at the innermost level of nucleons, it is crucial for the processes of fusion at issue here. In effect, weak interaction is what, at the very core of a nuclear structure, permits a proton to transform into a neutron. This process is known as *beta-plus decay*. Beta-plus decay enables the formation of the smallest nuclear structure, a deuterium nucleus comprising one proton and one neutron, at the moment when two protons collide at high speed.

For a simple but instructive image, picture three large forces at work when atomic nuclei and their "building blocks," nucleons, undergo modification. Electromagnetic interaction is what constitutes the ramparts around the "nuclear castle," repelling protons and other nuclei that fail to strike it at sufficient speed. In turn, strong interaction traps these particles when they do manage to pierce the wall, prompting "assemblages" of nucleons to form by confinement. Finally, within the castle, weak interaction can change the very nature of these nucleons, which is what enables the nucleus itself to transform.

This model contains the basic components of what French physicists Paul Langevin and Jean Perrin, Arthur Eddington in England, and other scientists had come to suspect toward the end of 1919. Researchers were trying to explain the immense energy emitted by stars, which are constituted by hydrogen clouds collapsing under their own weight, but their understanding of relativity and quantum physics was still piecemeal; thus, the existence of neutrons was not experimentally proven until 1932 (by James Chadwick, in England, who received a Nobel Prize for his achievement in 1935).

We should note that weak interaction, because it permits nucleons to mutate, is indirectly involved in the fission reactions of larger nuclei—and, more broadly, in the phenomenon of radiation. Even if the following chapters occasionally give the impression that (healthy) competition exists between partisans of fission and advocates of fusion, the science of physics reminds us that these processes are really two sides of the same coin.

Here, we have touched on the reason why fission and fusion are set in opposition to each other—or, more precisely, why a distinction is made between them. Nothing described until this point would a priori prevent the possibility of making large nuclei out of small ones, or breaking apart small nuclei in the same manner as large ones. Going a bit further, we can even affirm that nature itself has fashioned the hundreds of different nuclei on the periodic table—and therefore the world that surrounds us—in this very manner. How could so many elements have come into being in the first place without smaller nuclei fusing into larger ones? Indeed, following the initial phase of the universe's existence—the explosion known as the "Big Bang"—matter crystallized bit by bit out of the most elementary constituents. When the first nuclei took shape, the most numerous among them were hydrogen nuclei, which formed gigantic clouds. In turn, the gravitational collapse of these clouds caused the stars to emerge, which enabled hydrogen atoms to fuse and bring forth larger nuclei.

The reaction that makes nuclei smaller than iron atoms undergo fusion sets a great amount of energy free. Such energy is what maintains the heat necessary to maintain reactions and ensure the star's ongoing existence. This phenomenon, which is called *nucleosynthesis*, explains why nuclei lighter than iron are present in stars. That said, nuclei well beyond this *iron limit* may also be found. A second mechanism is at work, then. So-called neutron capture accounts for how stars create the vast array of nuclei one can observe in the universe as a whole. By "exploding," stars—which are mighty factories of matter—spread these nuclei abroad.

Here, we need to introduce a few new ideas about the energy balance in reactions of this kind. Once again, the fortress model offers a convenient and practical illustration. The height of the "castle walls" depends on the balance between electromagnetic and strong nuclear forces, and the area within them is suited for "holding" nucleons. Adding nucleons to the structure does not pose difficulty at the beginning, when there aren't too many of them. Indeed, such accretion occurs either through the process of nuclear fusion or through the neutron capture discussed above. But it is only logical that, as with any confined space, the castle will have a limited capacity to house new arrivals. It's not hard to imagine that it might overflow, thereby releasing nucleons or aggregates of nucleons. When the nucleus undergoes fission (whether spontaneously or by means of induction), that's precisely what happens.

We need to expand our somewhat simplistic model a little to account for further complexity. For one, the castle wall is not the same for neutrons, which have no electrical charge, as it is for protons, which do. Neutrons have an easier time getting through the wall and entering the castle. What's more, the wall does not block particles such as photons, which are susceptible neither to strong interaction nor to electromagnetic interaction. And the height of the wall changes in keeping with the relative fullness (or emptiness) of the castle, that is, the total charge of the nucleus and the number of nucleons present. Nor is that all. The wall is porous in the sense of quantum tunneling; in other words, particles can enter and leave with a kind of "passport," even when they don't necessarily have what it takes to jump out. Finally, "civil wars" rage here on the basis of abundant weak interaction and are constantly changing the nature of the population inside the castle; at any moment, the overall balance risks tipping over and putting the whole structure at risk. Somewhere between the worlds of *SimCity* and *Game of Thrones*, such conditions are fit to inspire video game designers of the twenty-first century—yet another surprise offered by nuclear physics!

For the purposes at hand, two major rules for global behavior within the atomic nucleus follow from these internal battles. The first concerns weak interaction, and its effects find direct expression in the periodic table. In a word, this rule holds that a valley of stability (cf. chapter 2) exists, which fixes the number of neutrons in relation to the number of protons under stable conditions. Any effort to alter this balance significantly yields unstable nuclei, which will disintegrate at different rates. Such radiation, as we have noted, occurs along three principal lines. Alpha radiation means that a helium nucleus is ejected; beta radiation entails either the ejection of an electron (this is beta-minus radiation) or the ejection of an antielectron, or positron (in the case of beta-plus radiation); finally, gamma radiation involves the ejection of a photon. The smallest nuclei on the periodic table are the most important for nuclear fusion. First among them is hydrogen; it has a single proton and is stable. The hydrogen isotope tritium has one proton and two neutrons, but its equilibrium is unstable; when it disintegrates, it emits an electron (weak interaction, beta radiation) and becomes a nucleus of helium-3, which comprises two protons and one neutron—and is stable. The time that such disintegration takes (its period, or half-life) is 12.32 years; this is how much time will pass before half of a given sample of tritium nuclei turns into helium-3. Tritium's half-life, while long enough for the element to be used for fuel in a fusion reactor, is much too short for it to be found in nature; tritium must therefore first be created (a point to which we will return). Finally, the fourth nucleus on the periodic table is helium-4, which is also called the *alpha particle*; it comprises two protons and two neutrons, and it is stable.

The second general rule concerns the energy balance of nuclear reactions themselves, which is tied to the physics of strong interaction. Quantifying the precise difference between the sum of masses of constituents in each nucleus and its real mass reveals a consistently positive, nonzero value. This "lack of mass" may be interpreted in terms of Einstein's famous equation, $E = mc^2$—which posits the equivalence between mass and energy—as the binding energy of the nucleus in question, that is, the energy required to break it down into its constituents (or, conversely, the energy obtained by assembling it out of separate nucleons). Tracing binding energy in reference to the number of nucleons for each nucleus in the valley of stability yields a concave bell curve; this curve grows very quickly for small nuclei, peaks at the iron nucleus (which comprises twenty-six protons and thirty neutrons), and then decreases slowly.

This result shows that a net gain of energy occurs when small nuclei assemble into larger and more stable nuclei, all the way up to Fe-56. The nuclear reactions at work transform elements with low binding energy per nucleon into elements with a higher binding energy per nucleon, which produces energy that can be recuperated. Such reactions are called *exoenergetic*. In contrast, a net loss of energy occurs if one tries to make "bricks" beyond the limit marked by iron. In this case, reactions are *endoenergetic*. A net loss of energy occurs if one breaks up a stable nucleus smaller than iron, and a net gain of energy is obtained by breaking a nucleus bigger in size. Here, then, we stand at the crossroads between two possible paths for producing nuclear energy. Fission exploits the net energy released by breaking up large nuclei, a process induced mainly by bombarding nuclei with nucleons. Fusion exploits the net energy released when small nuclei are formed by nucleons or nuclei that are even smaller. The general curvature described by nuclear binding energy also shows that if the energy released by fission already proves significant, the energy released by fusion is even more so—which makes it very appealing to the energy-hungry species that humankind has become in the course of evolution.

Nature "understands" as much. From a standpoint above the goings-on here on earth, it's clear that the universe as a whole is "heating itself" through fusion, and that it has been doing so for billions of years. To be more precise in chronological terms, the universe needed to reach sufficient spatial extension and then cool down, after the Big Bang, for basic matter to form: first elementary particles and then protons. At this point, the fundamental forces discussed above emerged in distinct form, which enabled the process of creation to unfold. Then, more time had to elapse for gravity—the fourth and final kind of fundamental interaction—to bring together immense hydrogen clouds, which eventually collapsed under their own weight and created the conditions necessary and sufficient for triggering the fusion reactions that "switched on," one by one, the hundreds of billions of galaxies constituting the universe we know.

The shift between fusion and fission explains the life and death of stars. In effect, the gravitational collapse of a hydrogen cloud triggers the fusion of nuclei it contains, thereby "lighting up" a star. In the course of a star's life, hydrogen is consumed, and new fusion reactions between nuclei already present enrich its composition with bigger and bigger nuclei. If nothing comes along to disturb the process of fusion at the star's core, it will eventually, after a few billion years, be composed of only iron nuclei. At this point, fusion stops and the star "goes out" by collapsing onto itself and then, in general, exploding. This is the case for a category of "large" stars with a mass eight to ten times that of our sun. In the phase before collapse, they are called *red giants* or *supergiants*; in the final, explosive phase, they are known as *supernovas*.

Humankind's adventure in fusion, on the modest scale imposed by the circumstances we have described above, is to try to reproduce—and control—the conditions necessary and sufficient for igniting "ministars" on our own planet. "Are you insane?" incredulous parties often ask when this venture is brought up. In a certain sense, we are. Seeking, like so many modern Prometheuses, to lay hold of divine fire does sound like a losing proposition. Astrophysicists are quick to urge discretion and modesty. Whereas our sun managed to get its nuclear power plant up and running four-and-a-half billion years ago, the planet Jupiter is there to remind us that the fire doesn't always take. The reason is quite simple: although this "planet" is really a huge ball of gas, it's too small to trigger the right kind of mutation! That is, the forces of gravitational compression at work are not great enough to produce the conditions under which hydrogen nuclei can overcome the Coulomb barrier and pierce the nuclear castle. Calculation tells us that Jupiter would need to be ten times bigger for this scenario to go into effect. To put things in perspective, any encyclopedia will tell you that Jupiter is by far the largest planet in our solar system, occupying a space one thousand three hundred times the size of the earth!

What is to be done, then? By what means can a star be "miniaturized" so it will serve human needs? In order to discuss fusion research scientifically and without polemics, we need to adopt precisely this vantage point. Whenever the first fusion reactor finally is built, it will amount to the greatest act of miniaturization human endeavor has ever produced. Including the attendant infrastructure, such a reactor will take up, at most, a few dozen hectares of land; its output will not exceed a few gigawatts—making it comparable to any other energy plant now in operation. These parameters represent the unavoidable anthropomorphic and societal constraints of fusion research. In terms of volume, the relation between one of the smallest stars—our own sun—and this human-made ministar will fall somewhere between twenty-four and thirty-three orders of magnitude[1] (for a reactor using magnetic confinement and one using inertial confinement, respectively)! Never, over the course of millennia of trying to imitate, reproduce, and surpass nature, has humankind faced such a daunting challenge. In comparison, early-modern voyages of discovery look like schoolyard games, and more recent projects to miniaturize technology seem like warm-up exercises for beginners.

Fusion research is not fascinating for scientific and technological reasons alone. It means embarking on a quasi-mythical journey to master the power of the stars. Perhaps the closest parallel is the conquest of space during the second half of the twentieth century: the richly symbolic "escape" from planet earth. For decades now, the fusion gambit has brought together a host of participants ranging from politicians and policy makers to field researchers and members of the general public. In equal measure, it has united dozens of professions and areas of scientific and technical specialization across cultural and geographical borders. It is clear, at present, that efforts will be complete more toward the end of the century than at its halfway point. There is little room for argument, then, that harnessing the power of fusion represents one of the greatest scientific undertakings in all of human history.

In this light, let's turn to the means that physics offers for miniaturizing a milieu that will be capable of triggering and sustaining nuclear fusion. The first step involves provoking a collision between two small nuclei with a velocity high enough for a reaction to occur. One or both of the nuclei must be accelerated and brought to a sufficient relative velocity. One way to do so is to aim a particle accelerator at a target. Alternatively, and more simply, a medium containing the atoms to be fused can be heated. To get an idea of the first option, picture a charge administered to deuterium gas. Doing so tears electrons away from one or more of the atoms and sets nuclei free. In turn, these nuclei may be accelerated by electrostatic means as they pass between two metal grids exhibiting different electric potentials. This creates a beam of "fast" deuterium, which can then be directed at a target with a high tritium content. To understand the second method, imagine a mixture of deuterium and tritium in a sealed container, which is "heated" to a level of thermal agitation high enough to trigger the desired reaction. In physics, *particle velocity* means kinetic energy, and the kinetic energy of a mass of particles is its *temperature* (to a close approximation). The key parameter, then, is the temperature of the medium in which fusion reactions are to occur. For the reaction between deuterium and tritium—the "easiest" collision, by nature—the necessary order of magnitude lies in the tens of millions of degrees (or higher). That's the first challenge.

But to get serious about energy production in economic terms, it's important to appreciate that the task confronting us is not to provoke just one fusion reaction, or even a few reactions; the goal is myriad reactions that will occur simultaneously and keep on going. Thus, it is not enough to ensure an appropriate temperature level in the medium containing the nuclei that are supposed to undergo fusion; it's just as vital to make sure that the number of reactions per unit of time is sufficient. Consequently, the number of nuclei per unit of volume—that is, the density of these same nuclei—provides the second major parameter for dimensioning the problem at hand.

Whichever path one takes, there is a third and final parameter that, like the other two, takes the collective and thermodynamic aspect of operations into account. Picture placing a pot of cold water on a burner, and then trying to bring it to a boil. Everyday experience teaches us that it only takes a few minutes to heat up a liter of water. But what happens when the amount of water is not one but ten liters, with the burner providing heat at the same level as before? Common sense tells us that the process will take longer, but we'll make it. Someone who's clever—and in a hurry—will get a lid from the cupboard and put it on top of the pot, knowing that this will speed things up. However, if we change the volume parameter even more and, say, put a hundred or a thousand liters in the pot, we can start to see that the amount of time required to boil the water will assume unreasonable proportions. Especially if the pot hasn't been covered—or if there's cold air coming through the window—conditions may not favor the water reaching the boiling point, even if we have all the patience in the world. This simple experiment brings us to the vital concept of *energy confinement* in systems in which a modification of temperature is desired.

Just as a bathtub will not fill unless the stopper is closed—or, at very least, something is present to slow the water from the faucet as it goes down the drain—no physical system can be heated unless energy is kept from leaving faster than it is added. The time it takes for a given medium to lose energy (or temperature) by a given factor (set at Euler's number, or about 2.72) without additional energy being provided is known as its *energy confinement time*. De facto, this value measures the quality of the medium's thermal isolation; the higher the energy confinement time, the better the medium is insulated from its surroundings. Increasing energy confinement time is how physicists "put a lid on the pot."

All the pieces are now in place to understand the basic, macroscopic criterion determining the possibility for a physical system to attain—and maintain—fusion reactions. The triple product of the temperature of the medium in question, its density, and its energy confinement time must reach or exceed a threshold value. Beyond this threshold, reactions will be sufficiently energetic and frequent for fusion, and the thermodynamic conditions of the medium adequate to maintain them. Credit for this discovery, made in 1955, goes to John Lawson, an English physicist who worked at Harwell Laboratory when fusion research was still classified. Ever since, scientists have spoken of the *Lawson criterion*. Besides honoring Lawson, this criterion points to one of the major paths to follow for solving the problem of miniaturization.

Even if it represents just one of the three parameters factoring into Lawson's "triple product," the temperature of the medium plays a key role because each fusion reaction has an effective cross section[2] (or range) around an optimal temperature, which the medium must reach. We must say goodbye to the idea of *cold fusion*,[3] then: density and confinement time cannot make up for the fact that nuclei need to get close enough to overcome the Coulomb barrier. That said, the other two parameters—density and energy confinement time—are a bit more flexible; this wiggle room has opened up research along different lines. In effect, one can try to maximize one parameter or the other, both separately and together. By compressing a mixture of deuterium and tritium, for instance, it's possible to trigger fusion reactions even if the medium (which is potentially quite small) presents a priori only a low energy confinement time. What's more, if arrangements are made so the medium's compression occurs very rapidly—in what is called an *adiabatic* process—it will trigger a simultaneous increase of density and temperature; the advantages for the problem at hand are clear. This approach is known as *inertial confinement fusion*. We will discuss this point at length below; for the moment, note that the most "miniaturizing" and self-evident means of obtaining fusion reactions is by rapidly compressing a tiny volume of mixed deuterium and tritium. The typical densities for such a mixture lie on the order of several hundreds of grams per cubic centimeter (in other words, several thousand times the density of the solid form), with energy confinement times of a few dozen picoseconds (a picosecond being one-millionth of one-millionth of a second). This is the principle underlying H-bombs. As we will see, domesticating—and therefore industrializing—a process like this poses real scientific and technical challenges.

In almost symmetrical, if not opposing, fashion, one can maximize the energy confinement time of the medium while preserving a reasonable—that is, a weak—density. A priori, this method holds greater promise when the task involves setting up a process maintained in stationary manner. That said, the size of the reactive medium poses a problem. To increase a given medium's energy confinement time, what's better than heating the medium while ensuring maximal insulation from the outside environment—as in our example of putting a lid on a pot or as one might do with a thermos bottle? Increasing the confinement time amounts to finding a "pot" capable of keeping a medium "warm," which has been brought to tens of millions of degrees (if not more), over a "long" period of time. Even in the best of cases, no material exists that will not melt or sublimate[4] above a few hundred or a few thousand degrees. Creating a high-performance environment requires real innovation. That's just one of the challenges fusion presents.

To understand how it is possible, all the same, let's look at heat on a microscopic level. One medium is said to be hotter than another if one of them transfers energy to the second when they come into contact.

From a microscopic point of view, such energy transfer can occur in two ways: by means of a flux of electromagnetic radiation or by a flow of particles. Thus, while the flame of a candle is hot because of the infrared radiation that it emits, it's also hot—and to a greater degree—because of the thermal agitation of gases generated by the chemical reaction of wax burning at hundreds of degrees (which rise above the flame). By placing one's hand to the side of, or above, the flame, it's easy to recognize the extremely high anisotropy of the heat that it emits; hereby, the exchange of heat between flame and hand passes through a stream of hot gas basically rising upward, which is compounded by isotropic electromagnetic radiation (which, in this particular case, is weaker).

In order to preserve the heat released by a given source, it's imperative to limit, block, and counteract streams of particles and radiation. Increasing the energy confinement time for any medium, then, means providing maximum isolation from the surroundings and, when possible, keeping fluxes of particles and radiation from getting outside in the first place.

It isn't easy to block photons or impede the motion of neutral particles like neutrons. De facto, it is only matter, with varying degrees of density, that can serve this purpose. In other words, we need to make a "bottle" around the reactive medium. Sufficient resistance must be in place for the neutral radiation (neutrons or photons) that is released to be absorbed by the material structure containing the reaction. As a rule, this enclosure is made of metal and needs to be cooled actively. The technology employed requires extensive testing to ensure that walls will be capable of collecting fluxes ranging from hundreds of thousands to millions of watts per square meter. This is the first constraint that applies to the global dimensioning of our bottle.

As in our example of the candle, there's another constraint, which surpasses the first and holds at a more local level. This is energy from the medium in fusion, which nuclei with temperatures from tens to hundreds of millions of degrees conduct directly to the vessel's walls. When they reach the sides of the enclosure, these energy flows must be contained; at the same time, they must be "slowed down" in order to increase the energy confinement time that they produce at the core of the medium, so it won't cool down too fast and bring the reaction to a halt. To this end, since the very first fusion experiments, a trick that capitalizes on a basic property of interaction between charged particles (be they positive or negative) and the magnetic field has been employed.

Any and every electrically charged particle that enters a magnetic field is subject to a force that makes it bend and describe a helicoidal trajectory. The pitch depends on the particle's speed along the magnetic field line; the radius of the helix depends both on the particle's speed perpendicular to the magnetic field line and on the intensity of the magnetic field itself. For a given particle of energy, the more intense the magnetic field is, the smaller the helix's radius will be—that is, the more the particle will have to "stick" to the magnetic field line, which traps it. At the same time, particles remain free to move along the field line, and they experience no constraint in this direction. In other words, if one manages to organize magnetic field lines that seal each other off, the result is an insulating magnetic space—a kind of "thermos" able to confine charged particles perfectly. As we will see, this model gets a little more complicated when put into practice, but the principle, which is quite simple and efficient, provides the basis for "magnetic confinement" in all fusion research—which we will now examine in greater detail. *Magnetic confinement fusion* is the term generally used for this method. The typical density for the mixture of deuterium and tritium lies at a few thousandths of a gram per cubic meter, offset by energy confinement times of a few seconds.

There are two paths for miniaturizing and domesticating stars, then. But what, apart from pure scientific inquiry—or mythological inspiration—is our motivation to do so? Why should we try to put a star in a box in order to obtain energy when other solutions exist that are simpler and easier? Without getting into debates concerning global energy transition, and leaving aside the evolutions and revolutions that occur almost every day on the level of energy sources, consumption, and planning, it's worth spending a little time on what fusion stands to offer. Deriving energy from fusion represents a radically new course in the project of harnessing nuclear power for civilian uses. Indeed, we can picture fusion as a complete break with the four generations of nuclear technology described up to this point.

The first advantage of fusion concerns fuel and waste. A fission reactor uses uranium (or another heavy element) and generates, in the course of operation, neutrons and radioactive products that need to be processed and stored. In contrast, a fusion reactor uses deuterium and tritium, and its reactions yield helium and neutrons. Deuterium is a stable isotope of hydrogen. Tritium, in turn, is a radioactive substance with a relatively short half-life, which must first be fabricated directly in the combustion chamber of the reactor (by means of the neutrons present). Various procedures now in development involve lining the reactor walls facing the mixture of deuterium and tritium undergoing fusion with material that has been enriched with lithium. Scientists speak of a "tritigenic blanket" that encloses the reactive medium. When a neutron leaving the medium collides with a lithium nucleus on the wall, it produces a tritium nucleus and a helium nucleus. Tritium nuclei created in this manner can be put straight back to use in the reactive medium. The idea is to produce and consume them "on-site," thereby avoiding the need to inject them from the outside or to deal with them as waste. From an external perspective, a fusion reactor will be fed with deuterium and lithium and release only helium—which harms neither human beings nor nature. The three elements in question are stable. Deuterium is found all over the earth (some thirty grams occur per cubic meter of seawater, for instance). Likewise, lithium is abundant in both the earth's crust and in the oceans; its concentration is estimated to lie at approximately twenty milligrams per kilo in the former, and at two hundred milligrams per cubic meter in the latter. Even if industries for extracting these elements do not yet exist, and in spite of the environmental costs associated with setting them up, the figures above indicate that natural resources for the fusion reactors of the future will be plentiful for tens of thousands of years—all but impossible to exhaust.

Apart from being abundant, deuterium and lithium are distributed uniformly across the globe. Unlike fossil and fissile fuels now in use, they are not scarce and therefore cannot give rise to geostrategic tensions. The new energy sector promises the immense merit of quasi-universal access to the necessary resources.

The second advantage concerns the "downstream" effects of the fuel cycle and what happens to plants no longer in operation. As we have seen, the fusion reaction itself doesn't generate radioactive output; this means that there is no need to create a secondary industry for recuperating, treating, or storing what plants produce. Nuclear reactions generate neutrons—and, in the case of fusion, particularly energetic neutrons that leave the reactive medium and hit the walls of reactors;

this slows them down, and then they are reabsorbed by the system. Even if most neutrons are reabsorbed by lithium in order to produce tritium in situ, it's important to note that the materials housing the reaction will be affected by them; potentially, the structure will be the site of induced fission reactions. This phenomenon, which is known to occur in plants already in operation, requires careful management from the outset, when the reactor is designed: strict controls govern the selection of building materials, and those that promote the production of radioactive waste (whatever its half-life) are avoided. One of the most well-known examples of such waste is cobalt-59; although this is a stable isotope used to make any number of steels, it will, when irradiated by neutrons, transform into cobalt-60, which is radioactive.

Proper selection of materials for fusion reactors requires research in its own right, then, and over the long term; it involves choosing elements resistant to neutron radiation along the lines described above, as well as further work on powder metallurgy, that is, understanding the behavior of crystal structures and developing suitable mechanical properties in materials that will be exposed to neutron radiation during the course of operations. In effect, neutron irradiation is known not just for changing the crystal structure of materials through atomic movement but also for generating helium microbubbles that can weaken the construction as a whole. The goal of research is to create, test, and implement models enabling plants to operate for decades on end (from fifty to eighty years, it is hoped); during this time, they should produce waste with very low radioactivity—without major or long-term consequences for processing or storage. Needless to say, materials will have to satisfy environmental and economic conditions before a viable and competitive industry can emerge.

Finally, the third advantage of fusion concerns the main point of criticism leveled at nuclear operations: safety. As we have seen, fusion reactions can occur only with sufficient heat in the medium and the triple product of temperature, density, and energy confinement above the threshold identified by Lawson. If these conditions are not met—or if, for one reason or another, they do not last—the reaction will go out on its own. Most important, reaction between the deuterium and tritium will not get started. Fission of a large nucleus induced by collision with a neutron generates smaller nuclei and one or more neutrons. In turn, the latter collide with heavy nuclei in the medium and provoke a chain reaction. More than one neutron has to be created by the reaction for the process to take off. If the level of neutrons at the core of the medium is not controlled by provisions for absorbing or slowing them, a fission reaction can become uncontrollable. Fusion between deuterium and tritium, in contrast, creates a neutron that does not enter into the next reaction; it escapes the medium without interacting with it. In consequence, the reaction process cannot run away—which guarantees this method's safety.

Comparing fission and fusion makes it clear that fission reactions are maintained by means of the neutrons they generate; since the core of a fission reactor requires tons of fuel to function normally for a period ranging from a few months to several years, *criticality accidents* can last for extended periods—and result in extremely serious problems, notably core meltdown. Fusion, on the other hand, does not involve a chain reaction; moreover, the chamber where reactions take place only ever contains a few grams of reagents, just enough to produce energy for a few seconds. Any incident in a reactor of this kind—including disruptions produced intentionally—would prompt the reactor to stop immediately.

Since the late 1950s, fusion has been recognized as a source of energy with sufficient merits to give rise to worldwide initiatives. Although, for strategic reasons, information was closely guarded at the end of the Second World War, the 1958 conference of the International Atomic Energy Agency in Geneva marked a turning point by declassifying magnetic fusion research. The domestication of energy obtained through fusion began—and with it a human endeavor on an unprecedented scale. Since then, debates on managing energy resources have stood front and center on the world stage, and often with a fair amount of international drama. More or less all conflicts in the second half of the twentieth century and at the beginning of the twenty-first may be viewed in light of struggles for access to energy. At the same time, societal constraints have come to weigh more and more on political efforts to resolve the attendant problems. Environmental considerations now factor into the celebrated "energy mix." This means looking for new sources of primary energy, as well as devising management solutions in the chain of production, distribution, storage, and consumption; the ingenuity and insight that are the order of the day concern technical matters and ecosystems in equal measure.

Fusion research—like nuclear power in general—doesn't involve the same time constants as energy from other sources, and the same holds for managing global networks effectively. Upper Corsica and the industrial region around Lyon, for instance, each require a different energy mix. By the same token, environmental factors are not identical for overdeveloped countries (such as our own, or China and India) and other parts of the world. Finally, it's clear that societies across the globe are consuming more and more electricity, and less energy from other vectors, which has had a marked effect on the sphere of transport. In this international picture, studies have shown a demand for massive amounts of electricity over the long term, and even the most optimistic projections for wind power or photovoltaic energy indicate the need for improvement on a planetary level. Research on nuclear energy is more than warranted, then, if electricity is to be provided on the appropriate scale without carbon emissions. As we will see, the scope of nuclear fusion research can seem small or large, depending on whether one takes a political viewpoint centered on financial considerations or adopts a sociological perspective focused on the evolution of human societies.

4
Pioneers of Fusion

> The most important thing for a genius is to be born at the right time.
>
> —Lev Andreevich Artsimovich

Let's go back to the 1958 Atoms for Peace conference. Even if the Americans, Russians, and British are the only ones with the technology, the hydrogen bomb is already a reality. New research can now focus on domesticating fusion in order to produce energy. That said, the situation will not be the same for magnetic confinement and inertial confinement. The latter will remain tied to the principle of the H-bomb for years to come; de facto, declassified research will concern only magnetic fusion.

This conference is the first global meeting devoted to nuclear energy for peaceful aims, a real turning point. More than five thousand participants are in attendance. Delegates express enthusiasm that research conducted in great secret—particularly by the triumvirate of the United States, Union of Soviet Socialist Republics, and United Kingdom—is now being shared, which augurs well for joint undertakings. Edward Teller, the Hungarian-American physicist acclaimed as the father of the hydrogen bomb, will later declare, "It is remarkable how closely parallel the developments in the different countries are and this, of course, is due to the fact that we all live in the same world and obey the same laws of nature."[1] But most striking of all is the impetus the conference stands to give to further research. After years of covert activity and strategic calculations against the backdrop of the Second World War, physicists from across the world have been invited to work together for a peaceable future that, it is hoped, will prove long-lasting. Discussing fusion in the Union of Soviet Socialist Republics, Lev Andreeevich Artsimovich, a visionary in the field and the keynote speaker at the event, stresses the importance of collaboration: "We must not underestimate the difficulties which will have to be overcome before we master thermonuclear fusion. . . . The solution . . . will require a maximum concentration of intellectual effort and the mobilization of very appreciable material facilities and complex apparatus. This problem seems to have been created especially for the purpose of developing close cooperation between the scientists and engineers of various countries."[2]

Time would prove the speaker right beyond anything he could possibly have imagined. Artsimovich, who was born in Moscow in 1909 and studied physics at Minsk, is our best guide to the pioneering phase of magnetic fusion. After taking up teaching duties—notably in Leningrad during the Second World War—he returned to Moscow in 1944 and joined the team of researchers working under Igor Kurchatov to develop an atomic bomb for Russia. In the two decades following the end of armed conflict, the Soviets stood at the forefront of efforts to meet the first challenge posed by magnetic fusion: finding a bottle that would contain a medium brought to tens of millions of degrees. (British efforts did not lag far behind.)

A whole book would scarcely suffice to describe the various paths researchers have taken over the years—much less assess their respective contributions. We must content ourselves with an approach that combines the overall history of these efforts and pedagogical explanation. The goal is to illustrate, as simply as possible, the course the scientific community followed to arrive at the two magnetic configurations that now represent the height of research.

Plasma

One must give honor where honor is due. In this spirit, we turn to the uncontested star of the fusion world: plasma, which scientific consensus deems the "fourth state of matter." In effect, the principle underlying fusion reactions is that deuterium and tritium particles must be able to collide freely: they need to draw close to each other at a distance on the order of their own size—just a few million billionths of a meter (or femtometers). For this to occur, the atoms must be stripped of their electron clouds, which otherwise hold them apart. The typical size of an atom is one hundred thousand times larger than the size of its nucleus; this means that nuclei which have not been completely "ionized"—that is, nuclei that still possess electrons—remain isolated, as it were, in an "electron cage" and cannot interact. To get a better idea of the distances at issue, picture the nucleus as a grain of quinoa at the middle of a football stadium; its electrons are tiny dots invisible to the naked eye, moving at the speed of thousands of kilometers per second, up in the stands.

Matter—whether constituted by atoms or molecules—changes state when heated. If solid when cold, it liquefies in response to heat. Its elements change position with respect to each other, or, more precisely, they slide over each other while remaining in contact. This is the liquid state. A medium's temperature represents the transcription, onto a macroscopic and collective level, of the individual energy of each of its constituents. If the liquid is subjected to further heat, it will vaporize—that is, turn into a gas. In this state, each atom (or molecule) frees itself from its neighbors. The medium no longer presents a regular structure but is instead the site of incessant and disordered motion on the part of its constituents.

Transition between solid, liquid, and gaseous states also depends on pressure. Some solids will turn into gases directly, without becoming liquids first. The term for this process is *sublimation*. The phenomenon takes place at temperatures ranging from "absolute zero" (–273.15°C, the lowest possible value, below which no movement can occur) to temperatures at thousands to tens of thousands of degrees; beyond this point, the thermal agitation of atoms reaches a level that permits collisions to wrest electrons from the electron cloud. Typical energy levels binding electrons to atoms belong to the order of tens of thousands of degrees. A gas heated to this point will begin to ionize and turn into a "soup" consisting of partially or entirely ionized atoms and electrons—all of which can move freely. This is plasma, the fourth state of matter. A given plasma counts as "cold" when atoms are not fully ionized, and "hot" when the soup consists only of electron nuclei that are completely separated from each other. Since the temperatures required for fusion reactions are extremely high, producing—and controlling—hot plasma poses quite the challenge.

In this context, let's return to the fusion pioneers and how to create a "magnetic bottle." At this early stage, researchers thought they were facing a fairly simple problem: within a plasma (nuclei and electrons in motion), the only relevant microscopic interactions are collisions, which occur at a frequency that depends on temperature and density. The magnetic field, reasoning held, should capture nuclei and electrons around its field lines; any energy loss would be caused by the collisions that regularly disturb the trajectories of particles. The bottle's dimensions would follow from the energy confinement time and the magnetic field for confining the plasma—all the rest should be a matter of finding the right geometrical proportions.

The Tokamak

The simplest design geometry involves a straightforward array of magnetic field lines, which are easily obtained by coils arranged in parallel to each other and perpendicular to the magnetic field to be induced.

This configuration yields a cylindrical form, a kind of pipe for confining the plasma (figure 4.1). The stronger the current in the coils is, the more intense the magnetic field will be—in other words, the greater the confinement of the plasma and its energy, a priori.

The first inconvenience arises right away, however. Ions and electrons escape at both ends of the magnetic "pipe" and prevent serious confinement from occurring. This is the case because the motion of charged particles experiences no restrictions in the direction parallel to the magnetic field. It's necessary, then, to close off both ends of the tube in order to obtain the desired effect. Bear in mind that our task is to create a trap, by

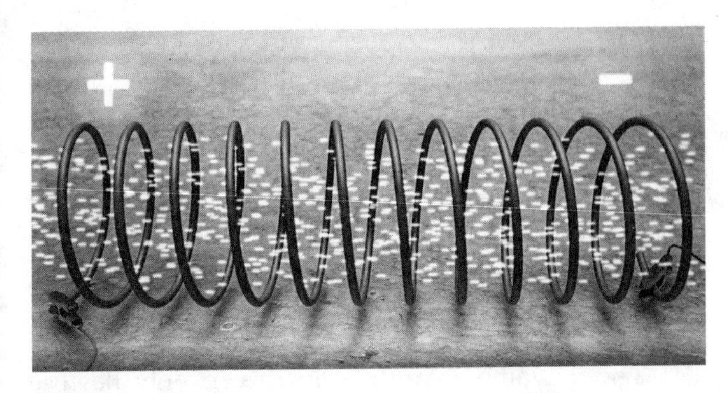

Figure 4.1
Magnetic field cylinder. *Source:* Commission for Atomic Energy and Alternative Energies (CEA), http://www.cea.fr/multimedia/Pages/videos/culture-scientifique/physique-chimie/fusions.aspx.

entirely magnetic means, which will be able to contain an extremely hot plasma. There can be no question of allowing the plasma to hit the material walls sealing off the ends of the magnetic cylinder directly. An initial solution involved heightening the intensity of the magnetic field at each extremity by increasing the current in the coils located there. This measure improved the situation greatly. In effect, it produced the phenomenon of reflection for some of the charged particles when they reached the outer parts of the tube, sending them back to the middle—in other words, "gradients" of the magnetic field acted like mirrors. That said, scientists soon recognized that this approach failed to confine all the particles. The "fastest" ones moving in the direction of magnetic field invariably escaped. The hotter the plasma, the greater the intensity of the magnetic field's gradient must be to achieve confinement. In consequence, the would-be solution was quickly abandoned. It simply isn't realistic to extrapolate these parameters to the scale necessary for nuclear fusion: there are too many technological limitations and problems with stability.

Efforts then turned to closing the cylinder in on itself by eliminating both "ends."

Magnetic volume confined in this manner has the form of a torus (figure 4.2); its large radius is determined by the distance between the central vertical axis and the horizontal circle running through the cylinder's center. This circle is called the *magnetic axis* (the arrows in figure 4.2). The *toroidal direction* is defined

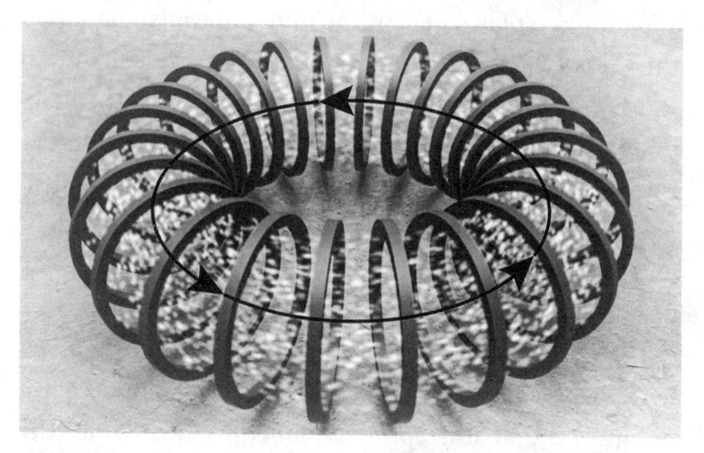

Figure 4.2
Magnetic torus. *Source:* Commission for Atomic Energy and Alternative Energies (CEA), http://www.cea.fr/multimedia/Pages/videos/culture-scientifique/physique-chimie/fusions.aspx.

as that of the magnetic axis, and the *equatorial plane* is the plane that contains the magnetic axis. Such geometry provides plenty of latitude for adjusting the large radius, on the one hand, and the vertical (or poloidal) section of the torus, on the other.

That said, a problem arises just as soon as one tries to contain a plasma: it escapes confinement under the effect of charged particles' vertical motion. Electrons and ions drift up and down in opposite directions. The effect derives from the curvature of the magnetic field, which exercises the equivalent of centrifugal force on the charged particles. The vertical separation of electrons and ions results in the loss of confinement, making this arrangement unsuited for the task. The major contribution of Soviet pioneers—Artsimovich, to be sure, but also Andrei Sakharov, Igor Tamm, and Igor Kurchatov—is to have devised an arrangement that offsets vertical drift while retaining the principle of toroidal configuration. If the field lines are wound in the form of helices around the magnetic axis, each charged particle will spend equal amounts of time traveling above and below the equatorial plane. Hereby, vertical drift is countered inasmuch as each particle is forced to reposition itself on the field line to which it is attached. In order to achieve this result, one need only add, at the outset, a component that turns on the plane vertical to the (toroidal) magnetic field. The resulting field line will wind around the magnetic axis like a helix (figure 4.3).

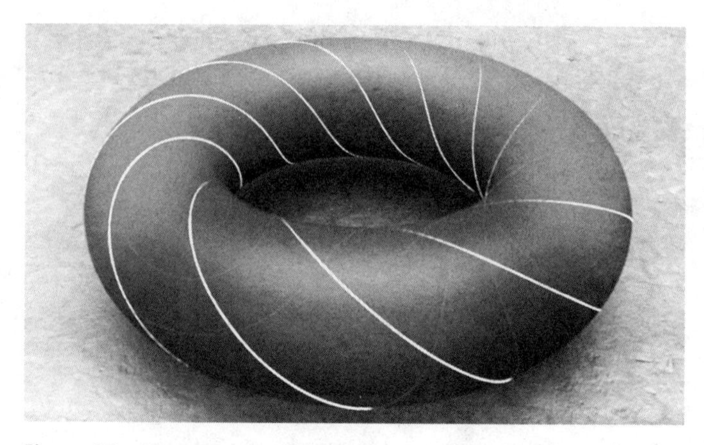

Figure 4.3
Coiled field lines. *Source:* Commission for Atomic Energy and Alternative Energies (CEA), http://www.cea.fr/multimedia/Pages/videos/culture-scientifique/physique-chimie/fusions.aspx.

There are two ways to create a so-called poloidal field in order to provide the necessary correction. Either one modifies the geometry of coils on the toroidal field, or—much more simply—one runs a current through the plasma in the toroidal direction. The whole secret of the тороидальная камера с магнитными катушками (in Roman characters, *toroidal'naja kamera c magnitnymi katushkami,* meaning "toroidal chamber with magnetic coils") lies in this solution: the celebrated "to.ka.mak" configuration (figure 4.4).

All the Soviet researchers had to do now was build the device. The first tokamak, in Moscow, was called T1 and had a toric chamber with a tiny volume: half a cubic meter of plasma. In 1968, the international community

Figure 4.4.
The tokamak configuration. *Source:* Commission for Atomic Energy and Alternative Energies (CEA), http://www.cea.fr/multimedia/Pages/videos/culture-scientifique/physique-chimie/fusions.aspx.

witnessed the sensational results of the T1 and its successors, T2 and T3. The temperature and confinement time of plasma in tokamaks far surpassed what other configurations, linear and otherwise, had achieved to date. The T3 even managed to reach ten million degrees. In 1969—at the height of the Cold War—a British team traveled to Moscow to verify Soviet claims. In consequence, research turned to magnetic fusion produced by tokamaks all across the globe. In France, to take just one example, the Atomic Energy Commission promptly constructed the country's first tokamak at Fontenay-aux-Roses. The TFR (*Tokamak de Fontenay-aux-Roses*), a design similar to the T3, went into operation in 1973 with a plasma current close to four hundred thousand amperes, which set a record for the day.

During the 1970s, more or less all the major fusion laboratories still active now were established. Sites include the Kurchatov Institute in Moscow, the Culham Centre for Fusion Energy near Oxford, operations at MIT, the Princeton Plasma Center, the Oak Ridge National Laboratory in Tennessee, and General Atomics, a private firm in San Diego. Japan, in turn, opened a center at Naka, in the Tokyo suburbs. Other examples include the National Institute for Nuclear Physics at Frascati (Italy) and the Max Planck Institute at Garching, near Munich (Germany). France founded two laboratories devoted to fusion under the aegis of the Atomic Energy Commission—besides the aforementioned facility at Fontenay-aux-Roses, facilities in Grenoble. Across Europe, partnership agreements between the Euratom program and member states multiplied and shaped fusion research for years to come. (The first contract was signed by France in 1959—some sixty years ago.)

Such massive expansion marked the end of the pioneering stage and the beginning of research on a grand scale, now focused on improving performance and surpassing standing limits. This work led to the great scaling laws in the field, that is, scientific understanding of how a magnetized, hot plasma's behavior depends on parameters imposed from without: geometry, volume, magnetic field, plasma current, density . . .

5
A Skillful Combination of Science and Technology

Before delving into the details of the second, and crucial, phase of research on the laws governing magnetic fusion, we should discuss the magnetic arrangement of the tokamak itself and understand its strengths and weaknesses as fully as possible.

Although the scientific community welcomed the new technology from the 1970s on, efforts have continued to document other magnetic configurations. The abiding goal is to use the knowledge acquired to make, at some point in the future, a reactor for producing electricity. The optimal magnetic configuration would ensure both stability and performance. As we have seen, tokamaks require that a current run through the plasma in the toroidal direction. In addition to promoting energy confinement, such a current offers the benefit of heating the plasma very efficiently by means of the simple Joule effect, since the plasma itself is a resistive medium (just like the heating resistance at work in an electric radiator). This added merit has helped the tokamak's fortunes, since it allows the device to reach record temperatures in little time. However, in order to create a poloidal magnetic field strong enough to contain the plasma, the current has to be quite intense. Moreover, inasmuch as the volume of the plasma is large, the stronger the toroidal magnetic field is, the stronger the plasma current has to be. Machines belonging to the generation of the T3 and the TFR used thousands of amperes; the next generation would need millions of amperes—with a view to a level around fifteen million amperes for the tokamak reactor of the future. Not only must this current be created but it must also be maintained. We will return to the methods that have been developed to this end; for the moment, the point to note is that the process consumes energy and, as such, affects overall efficiency. In concrete terms, it can take tens of millions of watts to maintain several million amperes in a plasma. This value represents a significant portion of the overall power one can expect to be produced by a fusion reactor.

Nor should we neglect other effects tied to the plasma current, notably considerations of the configuration's overall stability. Dozens of arrangements, mostly with names of poetic or exotic allure, have given way to the tokamak design—for instance, "magnetic mirrors," "zeta pinches," "theta pinches," "dense plasma focus," "reversed field pinches," "spheromaks," and "levitrons" (to say nothing of other "bumpy torii"). For all that, one other configuration is still receiving attention. It employs a magnet array similar to the tokamak, with helicoidal field lines running through nested magnetic surfaces, but it does so only with magnetic coils—that is, without a plasma current. This engineering coup is made possible by technology that gives a particular shape to the toroidal coils. The generic designation for this configuration is *stellarator*, and it continues to form the object of active research in Japan, Europe, the United States, and Russia. The stellarator has not taken the place of the tokamak—yet—for two main reasons. For one, the absence of a plasma current, and therefore of heating by the Joule effect, serves to reach plasma temperature. Second, designing and manufacturing induction coils that will create a magnetic field for a sufficient volume pose a problem beyond what is technologically possible at the moment. Still, considerable progress has been made in the last few years, both in terms of conceptualizing and creating a field at the necessary level and in terms of heating the plasma without a current. These advances account for renewed interest in the stellarator.

Thus, at Greifswald, in Pomerania, a new stellarator has just gone into operation. The Wendelstein 7-X (W7-X), which holds thirty cubic meters of plasma, is intended to display a performance level equivalent to a tokamak of the same size, with technology developed on the latter device. If results—which are expected within the next decade—should prove conclusive, they will mark a significant advance toward building a reactor. We can draw a parallel to the history of automobiles. The stellarator might be, with respect to the tokamak, what the electrical motor has been to the internal combustion engine. Although the basic design of cars stays the same, the efficiency of motors admits improvement. As we will see at the very end of the book, certain magnetic configurations "abandoned" in the past are also receiving renewed attention.

For the moment, we'll stick to the "royal road" of the tokamak and explore the scientific and technological path charted by the research community working on magnetic fusion during the 1970s. This second stage, which lasted until the beginning of the twenty-first century, focused on testing performance levels and limits. It consolidated the array of physical laws that account for the tokamak's operations on a macroscopic level as functions of its principal parameters: size, form, magnetic field, plasma current, plasma density, and so on. These laws, known as *scaling laws*, were deduced from experimental observation more than from theoretical physics. All the same, they have enabled engineers to envision improvements that have led to desired levels of performance and stability.

Creating a Plasma

The problem scientists faced as soon as they set about the first experiments on a tokamak involved knowing how to satisfy Lawson's criterion, that is, how to increase temperature, density, and energy confinement time simultaneously in order to reach the conditions favoring fusion. By the same token, they confronted the difficulty of ensuring that this plasma state, once achieved, would last. Improved designs facilitated researchers' understanding of the connections between various modifications and results. This is the thread tying together the matters discussed in the rest of the chapter.

A tokamak is a chamber in the form of a torus, surrounded by vertical coils distributed regularly in the toroidal direction. Once a magnetic field has been generated by the current circulating in these coils, the task is to create a plasma, which means ionizing a gas that has been injected into the chamber beforehand. To this end, it is imperative to empty the chamber of any other chemical compound by means of a high vacuum. *Vacuum chamber* is the term for the inside of a tokamak, the space where the plasma will be created and the process of fusion will occur. Prior to activation, the pressure level sought in a tokamak lies between one and ten millionths of one pascal—a value between one hundredth and one tenth of one billionth of atmospheric pressure. To get an idea of the scale in question, the pressure inside an incandescent light bulb is ten million times higher than in a tokamak, and the pressure of the interstellar void is one billion times weaker. The first difficulty is that this ultrahigh vacuum, without which the plasma cannot form, must prevail in the whole chamber—that is, in a volume of some dozens of cubic meters at present, which is anticipated to be as much as a thousand cubic meters in future reactors. Good old-fashioned mechanical pumps serve to reach a "primary" vacuum level, followed by so-called cryogenic pumps that further lower the pressure. The latter operate on the principle of trapping residual gas in the vacuum chamber on chilled surfaces with temperatures near absolute zero, where it condenses like steam on the sides of a refrigerator. Understandably, the process is relatively slow. In high-volume tokamaks, it can take several days, or even weeks, between the initial sealing of the chamber and the point where a suitable vacuum, ready for the plasma, has been achieved. Likewise, it's clear that ultrahigh vacuum conditions call for a mechanism that will ensure a perfect seal, notwithstanding the various parts that constitute it; otherwise, the necessary vacuum will never be obtained.

When these conditions have been satisfied, the vacuum chamber is ready for the gas, or mixture of gas, that will form the plasma. In a reactor, the gas will combine deuterium and tritium; for research experiments on smaller tokamaks, it's simply deuterium, hydrogen, or helium. For this gas to turn into plasma, it must be ionized. The most common method for doing so is to install electrodes in the enclosure with a current circulating between them. As in a fluorescent light bulb, high enough tension will prompt an electrical discharge that partially ionizes the gas and produces an initial "germ" of weakly ionized plasma. At the phase of "plasma breakdown," a cold plasma appears. Such "lightning" is the spark that sets the whole process into motion. In the interest of thoroughness, we should note that there's another way to trigger the ionization of a gas: injecting, from outside the torus, an electromagnetic wave at a very high frequency. This method is not in wide use, given the size of tokamaks at the moment, but it will become more and more relevant as plasma volume increases and the prospect of an actual reactor comes into view.

Generating the Current

At the moment of breakdown, the primary plasma, without any toroidal current, is neither hot nor confined. The few hundredths of a second that follow are crucial. Now, the central solenoid (figure 5.1) comes into play: a magnetic coil occupying the entire central hole of the toric chamber both in diameter and in height.

Before plasma breakdown, an electrical current is activated in this coil, saturating it with a magnetic flux that will be transferred to the nascent plasma by measured electrical discharges. In this way, the same relation is established between the central solenoid and the plasma ring as exists between the primary winding and the secondary winding in an ordinary electric transformer. Magnetic transfer from the central solenoid to the plasma turns into a difference of potential within the plasma. The effect of such electrical tension applied to the cold plasma ring is to accelerate electrons and

Figure 5.1
The central solenoid. *Source:* Commission for Atomic Energy and Alternative Energies (CEA), http://www.cea.fr/multimedia/Pages/videos/culture-scientifique/physique-chimie/fusions.aspx.

ions that have already been freed in the toroidal direction. Of course, ions and electrons are accelerated in opposite directions because they have opposite charges.

The desired current progressively emerges in the toroidal direction at a rhythm determined by the rate at which the central solenoid discharges. This is the phase of *current rise*. The usual rate of current rise in a tokamak lies between one hundred thousand and one million amperes per second; it takes between a few fractions of a second and a few dozen seconds, depending on the tokamak in question, to reach the appropriate plasma current. The emergence of the current is accompanied by that of the poloidal magnetic field; in this way, the overall magnetic configuration takes shape. At the beginning of this phase, the plasma is still relatively cold and resistive. As the current increases, the plasma ionizes in its entirety, easily reaching a temperature between a few million and tens of millions of degrees.

Creating a plasma current by discharging the central solenoid imposes a strict limit on the plasma's duration. Once the flux stored in the central solenoid has been used up, the plasma current will stop on its own. It is necessary, therefore, to find a substitute for this inductive mechanism in order to maintain the plasma current; otherwise, the plasma won't last long enough to be compatible with a reactor. We will return to this aspect (if not weakness) in due course.

To provide a complete picture, we should note the presence of one last set of magnetic coils. These *poloidal* coils surround the tokamak horizontally (figure 5.2),

Figure 5.2
Poloidal magnetic coils. *Source:* Commission for Atomic Energy and Alternative Energies (CEA), http://www.cea.fr/multimedia/Pages/videos/culture-scientifique/physique-chimie/fusions.aspx.

and their main role is to permit control of the plasma's vertical section and global stability.

Heating the Plasma

Once the current plateau has been reached—or sometimes even beforehand—a saturation of the plasma temperature sets in; however, it occurs at levels that cannot possibly satisfy Lawson's criterion. This limitation derives from the inverse dependency of the plasma's resistivity to its temperature. The matter is easy to understand if one bears in mind that the hotter the plasma is, the more freely ions and electrons are able to move within it, which weakens its resistance to an electric current. In order to raise the temperature and reach the hundreds of millions of degrees necessary for fusion, a means must be found to continue heating the plasma now that it is both hot and confined, but in a way that does not rely on induction—that is, without relying exclusively on the Joule effect.

Just as temperature in a pot can be raised by pouring hot water into cold water, the temperature of a plasma can be increased by adding more energetic particles: energy is redistributed when these superheated particles collide with those of the plasma. The process that corresponds to making water tepid is called *equipartition.* This first way to heat plasma, then, is to accelerate the deuterium and tritium nuclei outside the vacuum chamber by traditional electrostatic means and to inject them through an aperture in the vacuum chamber (figure 5.3).

In practical terms, however, such an injection system poses an array of questions and challenges. The first—and biggest—difficulty is that the tokamak configuration has been designed to prevent charged particles from exiting (or, at least, seeks to keep them to a minimum); as such, it's extremely hard to inject them from outside. To be able to enter the plasma, these ions, which are highly energetic, must be provided in the form of energetic but neutral atoms, which subsequently are ionized by collision and transfer their energy to other particles of the plasma.

In brief, the challenge involves creating beams of fast neutral atoms in external equipment attached to

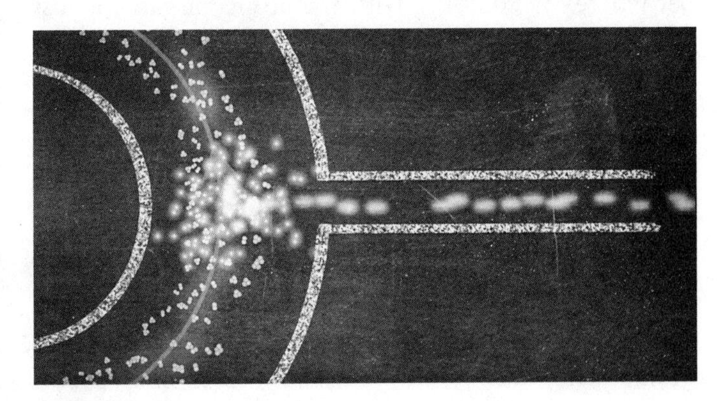

Figure 5.3
Additional plasma heating by particle injection. *Source:* Commission for Atomic Energy and Alternative Energies (CEA), http://www.cea.fr/multimedia/Pages/videos/culture-scientifique/physique-chimie/fusions.aspx.

the vacuum chamber. Of course, the first technology to have been adapted for tokamaks and stellarators was inspired by the ion accelerators in high energy physics laboratories. Such systems first create a cold plasma with elements ionized to varying degrees in a chamber where electrostatic charges are administered (the process is similar to the technique of initial plasma breakdown). The resulting ions are positive. For the case at hand, the gas is deuterium or tritium, so it's fairly easy to strip the atoms of their sole electron; the resulting deuterium or tritium nuclei can then be accelerated between the electrodes of an acceleration chamber—a technology that now is highly developed. The beam of fast nuclei obtained must then cross a cloud of deuterium or tritium gas in a so-called neutralization chamber where, by colliding with the atoms of gas, the fast ions recapture an electron and go on their way, now as fast neutral atoms. At this juncture, the beam passes through an aperture in the vacuum chamber and then heats the plasma.

There's no need to get bogged down with the sophisticated technical and physical details of such a system; suffice it to say that neutral beam injectors are now in place in more or less all the tokamaks worldwide. Each beam carries between one million and several million watts, and the ions in the beam have an energy level two to ten times higher than the temperature to be achieved in the plasma. Typically, ions are accelerated to some hundred thousand volts, which yields beams of high intensity. Although the state of research in the field is quite advanced, use of these elaborate electrotechnical systems is restricted to the world of magnetic fusion. As progress continues toward better control of fusion and, ultimately, a functioning reactor, the amount of energy injected will only increase. As the plasma gets larger, hotter, and denser, fast ion beams with the above characteristics will penetrate the plasma less and less. Deposited only at the periphery, their energy will not contribute as much to the desired heating of the plasma core. Logically, then, bigger, hotter, and denser plasma requires higher-energy beams. The problem is that, even if technology permits us to increase the electrostatic acceleration of ions within the injector, the effective neutralization of the beam diminishes quite rapidly as soon as the level of one hundred thousand volts typically used in research tokamaks has been exceeded.

Here, science confronts a pretty stubborn physical limit. The technology we've been describing cannot be expanded to the size necessary for a fusion reactor. The solution now in development addresses the problem at the level of the ions' source. It involves creating an injector that starts with deuterium or tritium gas which is ionized negatively—in contrast to the process of positive ionization, described above. The principle is similar, but with the difference that the first stage no longer uses a plasma of (positively charged) deuterium or tritium; instead, the objective is to attach an electron to deuterium or tritium atoms in order to produce cold ions with a negative charge. In turn, these negative ions are accelerated; when, at a high level of energy, they enter the neutralization chamber, collision with the neutral gas will tear away the supplementary electron, thereby releasing fast neutral atoms to much greater effect.

While simple on paper, the procedure isn't so simple in practice. Much of the difficulty stems from creating negative ions on a massive scale. It can be done in chambers where a cold plasma of positive ions and atoms—which is easily obtained—interacts with walls made from appropriate materials. The physicochemical exchanges that occur on this surface create negative ions from the positive ions that strike it, with a low but altogether acceptable output. Describing the particulars of this process would lead us too far afield; what counts is the fact that the phenomenon occurs—and that it can be optimized by adjusting the composition of walls and the energy level of plasma ions hitting them.

Once again, our kitchens offer a way to understand the relevant method for transferring energy from one medium to another. Instead of adding a hot liquid to a cold one to raise the temperature, why not just put the cold water in an oven (either of the traditional sort or a microwave)? As surprising as it sounds, this is basically what is done to heat tokamaks, and the practice was in place even before microwaves became a feature of most homes. As before, the principle is simple, even if practical implementation proves to be more complicated. If one or more of the waves at the periphery of the plasma in a tokamak get "excited" and propagate to the core, they can deposit energy there (figure 5.4).

In ovens at home, the molecules of water present in food absorb the energy conveyed by infrared waves or microwaves. Plasma, for its part, is made up of charged particles—ions and electrons—which are sensitive not just to static electrical fields and magnetic fields but also to oscillating electromagnetic fields of other kinds, that is, to waves in general. It so happens that electromagnetic waves—for instance, those of old-fashioned radios, the x-rays used in medicine, and ordinary light—all consist of combined electrical and magnetic fields in oscillation propagating through space at the speed of light. These waves spread more slowly in a medium that is not a void—say, water, air, or glass. The *refraction index*

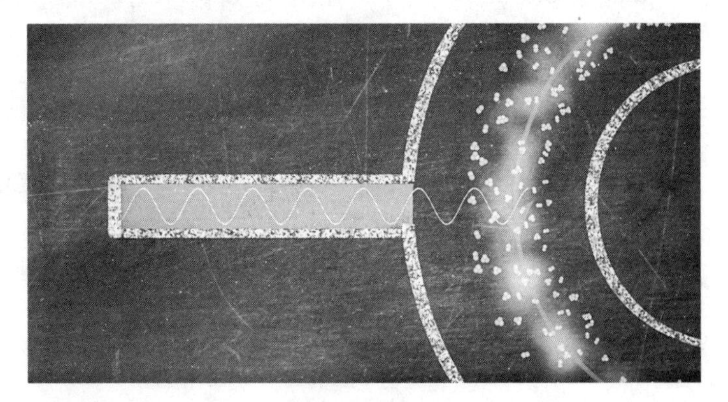

Figure 5.4

Additional heating of a plasma by wave injection. *Source:* Commission for Atomic Energy and Alternative Energies (CEA), http://www.cea.fr/multimedia/Pages/videos/culture-scientifique/physique-chimie/fusions.aspx.

describes the relation between the speed of light in a void and its speed in a given medium. This value is a dimensionless number greater than one. If a medium is said to be optically active or nonlinear—like a hot plasma—the refraction index proves to be much more *complex,* in keeping with a host of parameters concerning the wave and the plasma itself. Hereby, the wave's propagation is much more involved than is the case for a regular beam of light; various components of the electrical and magnetic fields play a role.

In spite of these additional factors, the picture of the process is clear enough to predict results with relative precision. After putting the propagation of an electromagnetic wave between the periphery and the core of the plasma into the equation, one must find the means, singular or plural, to get the wave to deposit its energy there; this should occur as efficiently as possible, which means adjusting the various parameters. The main, and most effective, way for a wave to transfer its energy to a charged particle in a magnetic plasma may be understood in analogy to a mother pushing her child on a swing! This simple example illustrates the physical phenomenon of *resonance* between the periodic movements of a source and a target. If a tiny push occurs "at the right time," the mother will transfer her energy without any difficulty, and the swing will go higher.

The same thing happens with a wave and hot, magnetized plasma. When the wave "pushes" the ions or electrons at the same frequency as the natural frequency exhibited by their motion at the plasma core, energy is transferred. Scientists have already discovered at least two periodic motions displayed by charged particles at the plasma core. The first concerns particles trapped around the magnetic field lines (*cyclotronic* motion, which is perpendicular to the magnetic field). The second (*parallel* motion) involves particles traveling down these same magnetic field lines; this motion is periodic because the magnetic field lines are coiled.

When discussing the problem of energy transfer between waves and plasma, we need to bear in mind that there are two types of particles, ions and electrons, and two types of periodic motion, parallel and perpendicular. The frequencies typical of the cyclotronic motion displayed by ions in a fusion plasma, which depend on the local magnetic field, belong to the order of several dozen megahertz, that is, a few dozen million oscillations per second. Such frequencies are quite close to those of FM radios. The technology for wave generators at these frequencies is well-known. The devices are high-frequency amplification tubes called *tetrodes*. As a rule, tokamaks require between fifty and one hundred megahertz and between one and two million continuous watts per tube. The electromagnetic excitation generated by this means is transported from the generator to the vacuum chamber by coaxial cables of the same kind as standard television cables, adjusted for size. It is then "injected" into an antenna facing the plasma, which serves to transform incident power into a wave that can propagate to the plasma core, where conditions of resonance with the ions prevail. Here, ions at the desired resonance level absorb it, and their own energy increases markedly. Thanks to collisional equipartition with the other particles, the plasma as a whole is heated.

In similar fashion, it's possible to link waves to the cyclotronic motion of electrons in the plasma. Doing so requires a wave much higher in frequency, one in the range of one hundred gigahertz. At such frequencies, scientists enlist the technology of microwave oscillators, which work by transforming the energy contained in an electron beam into the right kind of wave by mechanical means that impart the desired characteristics. These oscillators are called *gyrotrons*. The wave generated is transported from the gyrotron to the tokamak in the same way that a beam of light would be, via metal waveguides that lead to an opening in front of the plasma. The wave can now couple with the plasma without further modification. It's also possible to direct the wave by means of a simple arrangement of adjustable mirrors. The wave's absorption by electrons in the plasma that resonate with it causes the plasma to heat up. Gyrotron technology is fairly widespread in the industrial sector, but the particular demands of fusion (frequency, power, and duration) have long posed a challenge. Only in the last few years have industrial gyrotrons been made—in Russia, Japan, and Europe—that satisfy these requirements. We should note that researchers in China have been showing interest, too, and may soon announce revolutionary advances.

Finally, a third kind of wave is used to heat tokamaks, but in a rather different way. The goal is to accelerate electrons directly in parallel to magnetic field lines. Resonance is set at the level of an electron's periodic passage in front of an antenna that has been installed for this purpose. The frequency of motion falls in the range of a few gigahertz. Consequently, this method requires enlisting another kind of high-frequency tube, the so-called klystron. Such technology is well understood, even though, once again, fusion places demands of power and duration that require further research and development. Transporting power from the generator and the vacuum chamber occurs by means of waveguides, but to obtain the desired effect on the plasma, special antennas must be installed. In effect, the wave injected into the plasma has to be configured so that electrons will accelerate in just one direction, in keeping with the wave's toroidal propagation. The antenna takes care of that. The technology employed here consists of rows of waveguides that open onto the plasma at a few centimeters' distance. At the opening of this "grill" (which is what the antenna looks like), individual waves combine into a single wave displaying the desired properties.

The reason for accelerating electrons in one direction is to maintain the plasma current without consuming the magnetic flux from the central solenoid. Here, we can glimpse a partial solution to the problem of the tokamak's transitory character. Surpassing this limit would open the way for continuous operation, that is, for ongoing energy production. The process belongs to an array of scientific solutions called *noninductive current generation*. The other methods of added heating we have noted also offer advantages in this domain. Dissymmetrizing the waves injected and steering beams of neutral particles in the toroidal direction are ways of producing effects of noninductive current generation. The last two decades have witnessed major international efforts to maximize noninductive current generation by adding heat; the goal is to apply advances to future reactors. That said, the basic physics governing both the interaction between waves and plasmas and the transfer of motion generated by collision stands in the way of satisfactory results in terms of efficiency. At the moment, no solution exists for maintaining the fifteen to twenty million amperes necessary for a reactor—at any rate, not without expending more energy than will be produced. The problem is that the efficiency of machines of the current generation, defined in terms of the ratio between current generated and power expended, is at least one order of magnitude too small.

Even though the physics of the interaction between wave and plasma is disappointing on this score, collective work on the physics of the plasma itself offers a pleasant surprise: the *bootstrap current*. In effect, a hot plasma will generate a substantial current on its own when it reaches a high level of internal pressure. Without getting into the details, let us simply note that, under the pressure conditions needed for the fusion of deuterium and tritium nuclei, a bootstrap current can represent a sizable portion—if not the larger fraction—of the requisite plasma current. Once optimized and combined with noninductive current generation, it opens the way for the tokamak to function continuously. Research on this operating regime is not yet complete. If only for the sake of anecdote, we should remember that this effect was given its name with a nod to the famous Baron Munchausen, who managed to fly by pulling himself up by his bootstraps.

We can now sum up the strategic and critical challenges posed by added heating and generating fusion plasmas by noninductive means. The elements of the physics involved are well understood, and effective models are in place; likewise, reliable methods have been developed for heating fusion plasmas to the temperatures necessary for triggering reactions. To maximize the output of future reactors, progress must yet be made in terms of injecting neutral particles, developing industrially reliable solutions over the long term for all systems, and improving the overall balance of energy. Last, but not least, we should bear in mind that, for basic reasons of efficiency, noninductive current generation in a tokamak cannot rely on simply adding power from outside, no matter the source. It's important for a significant portion of the current to be self-generated by the plasma.

6 A Science of Compromise and Control

Seeking and Recognizing Performance

We have reached the end of the *preforming stage*: a plasma current has been established at the target value, the plasma itself is completely ionized, and the deuterium and tritium nuclei are confined along the magnetic field lines and able to collide under conditions favoring their fusion. So what happens to the plasma and its environment now? What do we "see," in concrete terms? What determines whether all this will "work" or not?

Although naive, these questions are wholly legitimate, and they help us appreciate the distance still separating us from our ultimate goal. A fuller picture of the physics involved will help us draw closer.

First of all, we should remember that if density and temperature represent crucial components of Lawson's criterion, they aren't the only factors. Energy confinement ensured by the magnetic configuration provides the third key we need to open the door to fusion. We now turn to aspects of plasma physics central to how the medium confines its energy. Fortunately, the main provision is already in place: the tokamak, or "magnetic bottle," which ensures that charged particles will follow a trajectory buckling back on itself indefinitely, provided that nothing comes along to bother them. The problem is, a few troublemakers get in the way . . .

The first of these unwanted guests appear among the collisions between particles—in spite of the fact that collision is vital for fusion to take place at all. What, exactly, happens when two charged and confined particles in a plasma collide? Whether or not fusion takes place, the particles' trajectories are affected in keeping with the universal (and fairly simple) laws of the conservation of energy and motion. These laws allow for the extremely free transfer of velocities between particles parallel and perpendicular to the magnetic field, as well as shifts of position, however tiny. Just picture two billiard balls hitting each other to get an idea of the disruptions at work. Every collision upsets the parameters of the trajectory for each of the particles involved. While it's possible to model this entire process, what scientists want to know is the global effect on all the trajectories of all the particles for all the collisions occurring at every moment in the medium. The plasma density in a tokamak is weak, but it still lies on the order of 10^{20} (one hundred billion billion) particles per cubic meter.

Statistical physics tells us about the twofold effect that takes place. The first effect is equipartition, which we discussed above: if the kinds of particles present in a collisional medium have different energy levels, both time and the collisions themselves will take care of uniformizing them through diffusion, making populations evolve toward identical temperatures. The second effect, however, is that these same populations will tend toward uniform distribution in space, chasing charged particles away from the plasma core toward the periphery; in other words, the plasma energy will tend to become unconfined. Such motion, which is also a form of diffusion, may be pictured in analogy to putting a drop of syrup in a glass of water—it won't stay put. The problem's clear: the more we try to promote fusion reactions in the plasma core and the more we try to maximize collisions, the more pronounced the collisional mechanism will become—and thwart our efforts. The balance the medium achieves between magnetic confinement and the deconfinement due to collisions is called *neoclassical equilibrium*. The equations accounting for this phenomenon are now sufficiently established that we can make precise predictions about the shape that plasma confinement should take in keeping with the macroscopic parameters of the tokamak and the plasma itself.

It is possible, then, to calculate and predict the neoclassical confinement time of a plasma with reasonable precision. Indeed, calculations come close enough that scientists were able, early on, to compare projections with what they observed on tokamaks in operation.

The results, in the 1970s and 1980s, produced the first major psychological setback that researchers faced: as it turned out, theoretical predictions and experimental observations didn't agree at all. Worse still, the facts offered far less reason for hope than the neoclassical model had suggested. In particular, predictions about the interaction of density, temperature, and magnetic field didn't agree with what actually occurred and signally failed to explain the experimental tendencies in evidence. The more the fusion plasma was heated, the more the energy confinement time deteriorated. Apparently, another physical phenomenon was occurring instead of the plasma energy (generated by collisions) simply spreading to the exterior as expected. Far from it: neoclassical theory held that a tokamak the size of the Joint European Torus (JET), operated by the European Union at Culham in England—one hundred cubic meters of plasma confined by a magnetic field between 3 and 4 teslas[1]—would easily prompt plasma composed of deuterium/tritium to ignite—that is, allow its nuclei to fuse in a self-sustaining manner, without the addition of supplementary power.

Once again, the phenomenon may be understood in analogy to a pot of water. In effect, heating the pot's contents by adding hotter water or by lighting the burner aren't the same thing. In the first case, the hot water diffuses in the cold water; in the second, heat from the flame causes local overheating, which then spreads to higher and colder levels through vortices in the liquid. These convection vortices are easily observed, and the answer to our enigma lies with them—a feature the first models left aside. That said, our analogy with the pot of water stops here. Scientists duly noted the vortex phenomenon (which is more appropriately designated by the generic term of *turbulence* in physics) in plasma constituted by fluid media, both liquid and gas. However, it took time to identify its properties and develop adequate predictive models; nor is the issue entirely resolved, even now. A fluid medium, with high or low viscosity, starts out in a homogeneous state; in response to a heat source, the addition of material, rotation, or some other physical quantity to which it is sensitive, it then becomes nonhomogeneous. The excitation that occurs causes it to modify one characteristic or another—temperature, density, speed, and so forth. Within the medium, a geographical difference, also known as a *gradient*, emerges. The gradient of a given physical quantity is its variation per unit of distance (a temperature gradient is expressed in degrees per meter, for example). Turbulence originates and draws its energy from such gradients. From this point forward, it will flatten out the same gradients that produce it—that is, it will make the medium in which it evolves return to a homogeneous state. In this respect, turbulence has the same "goal" as collisions, although this occurs by way of a much more "efficient" process: generating vortices of varying dimensions in order to homogenize the medium as fast as possible.

Every medium has a corresponding mechanism of turbulence, as does every gradient. Here lie, to this very day, the worst nightmares for fusion researchers trying to devise models of the microscopic physical processes underlying the generation of vortices in magnetized plasmas. A given medium is quite complex, especially inasmuch as the presence of a magnetic field makes it anisotropic. It's not easy to find a simplified scheme for the motion of ions or electrons; all the details of space and velocity must be considered. Coming up with a model for the initial appearance and subsequent development of turbulence means solving numerical problems at the outer limit of what is now possible. The most powerful computers aren't up to the task; even state-of-the-art machines can only work in overly simple terms—on a small scale, with brief temporal intervals much lower than the energy confinement time actually needed, or by focusing on just one kind of turbulence in the plasma. Advances in computing technology give reason to hope that researchers will be able to simulate the development of turbulence in full in the next ten to twenty years, prior to the construction and launch of a fusion reactor.

In the meanwhile, what do we know about our boisterous guest? What influence does turbulence really have on energy confinement time? Above all, turbulence is what determines and dominates the transfer of energy in a tokamak from the plasma core to the periphery. The heat generated by fusion reactions in the innermost—that is, the hottest and densest—part of the plasma causes a temperature gradient to form, which creates turbulence and thus convective movement toward the plasma's periphery. Such "boiling" will hardly surprise anyone who has seen images of the sun's surface—especially the spectacular recordings made by the Solar and Heliospheric Observatory satellite a few years ago. This basic picture is accurate, but there are other elements at work behind the scenes. Gradients of temperature, density, and global plasma rotation are also present, generating not one but multiple sites of turbulence, each on a different temporal and spatial scale—which makes the real situation even harder to analyze in detail.

For want of exhaustive modelization, fusion researchers have, for more than two decades, been devoting a major portion of experiments to documenting phenomena of turbulent transport and energy/particle confinement. The amount of information that has been gathered is considerable, and cooperation across borders is intensive and well organized. Studies have made a wealth of data available and led, in particular, to the scaling law that governs energy confinement time in tokamaks.

These parameters are the characteristic magnitudes of a given tokamak's geometry, as well as the values that can be assigned, when conducting experiments, to the magnetic field of confinement, the plasma current, average density, and additional heating levels. Scaling law doesn't automatically imply a physical model, nor does it explain underlying phenomena. It's simply the result of mathematically interpolating a very large number of results obtained while varying parameters in an experimental setting. As such, it serves as a guide when designing new tokamaks and verifying parameters of size, magnetic field, or current. In particular, the scaling law concerning energy confinement time represents one of the most fully documented aspects of magnetic fusion, and it teaches us many important things. For one, it confirms that plasma volume plays a crucial role. It's not hard to understand why the turbulent nature of heat transfer described above is a point of reference. The "bigger" the plasma is in terms of turbulent elements, the better its isolation from the surroundings—which means that it will retain heat that much longer. This consideration accounts for the projected size of fusion reactors, which we will discuss in the following chapters. Scaling law also confirms that the plasma current and the magnetic field affect confinement directly, and in concert. Finally, and with a relatively high degree of precision, it quantifies how energy confinement deteriorates with heating power—a macroscopic phenomenon expressing the presence of turbulence within the plasma.

As we've seen, heat propagates through turbulence, progressively extending from the plasma core to its periphery. At the core, temperatures of hundreds of millions of degrees prevail; the plasma grows cooler as it approaches the material structure forming the vacuum chamber. At the plasma core, then, a so-called equilibrium temperature holds; the overall temperature profile displays a bell-like shape as levels decrease to either side. If the plasma core is "overheated," its first response is to generate more turbulence to prevent its temperature profile from changing. This phenomenon is the deterioration of confinement time with power, as noted above. Surprisingly, however, there are situations in which increasing power even more prompts the plasma to shift abruptly toward a different state (or different states) of confinement, which allows it to get much hotter and, therefore, store more energy than anticipated.

Such bifurcations in heat confinement are tied to the spontaneous emergence of areas in the plasma where turbulence suddenly drops off, or vanishes entirely. In this context, researchers speak of *transport barriers*: the turbulence spreading heat is literally blocked, with the result that the net storage of energy in the plasma improves. This improvement on a microscopic level translates to the macroscopic level, in keeping with scaling law, by a numerical coefficient christened the *H-factor*. Two major types of barriers can be observed experimentally. The first barrier emerges at the outer edge of the plasma, appearing almost automatically beyond a critical level of additional power. It "insulates" practically all of the core plasma from the periphery and constitutes a "pedestal" to which the plasma's global temperature profile then rises. The factor by which energy is stored is considerable, on the order of two. This improved state of plasma confinement, discovered in the 1980s on the German ASDEX (axially symmetric divertor experiment), has thenceforth borne the name of *H-mode*. In the intervening decades, it has been reproduced and characterized by observation on more or less all tokamaks in operation; now, it provides the point of reference for all new machines—including reactors yet to be built. The second type of barrier does not appear so much in the periphery of the plasma; in general, it insulates a very small volume around the plasma core. Its trigger factor is less well-known and not just a matter of the level of additional power. This barrier is more closely tied to the local values of parameters and more difficult to manage. In effect, all parameters act on points of turbulence and can, in some cases, trigger local stabilization. Since the volume affected is less significant with this second kind of barrier, the corresponding factor of improvement for confinement generally proves to be lower; all the same, it has a bearing on fusion reactions. We should bear in mind that the self-generated "bootstrap" current feeds off all improvements to pressure and its gradients; that is, it benefits from these same barriers. Internal barriers are still being studied; they have not yet been well enough understood to play an integral role in reactor design. Finally, we should note that both kinds of barrier have no trouble coexisting—which enhances opportunities for further observation and improvement. *Advanced tokamak* is the generic title applied to this branch of research.

As remarked above, theoretical and numerical descriptions of hot, magnetized plasma represent a highly technical aspect of fusion research. Advances in knowledge and prediction are closely tied to the capacities of powerful computers. Today, no computer exists that would allow us to calculate the turbulence that develops and produces the confinement times that have been observed or the bifurcations we have been discussing, especially in the case of large tokamaks. It will take a few more generations of computer technology—and maybe a few more generations of physicists—before the next stage is reached. All the same, descriptions and calculations already available allow us to affirm the key role played by turbulence and to shed light on its principal effects. Such information also inspires confidence that steps are being made toward building a reactor.

Ensuring the Plasma's Stability

Everything would turn out well in this best of turbulent worlds if efforts to achieve optimal plasma performance were not limited by nature—to wit, by certain fundamental mechanisms at work involving currents, pressures, and magnetic fields. We can approach this vexed field of endeavor by calling to mind a few scenes familiar from high-school science classes.

By definition, fusion plasma needs to be kept in a state of constant "nonequilibrium" because the point is to generate energy at the core, which then passes to the periphery. Over and above the "continuous" phenomena that take care of regulating fluxes of heat, particles, or any other quantity in a state of disequilibrium—collisions and turbulence, for instance—sometimes it is possible to observe phenomena that are much less calm or regular but serve the same function in a less gentle manner. To get back to our pot in the kitchen, convection takes care of mixing the hotter water at the bottom with the colder water on the surface, so that local temperature never exceeds 100°C. As soon as this temperature has been reached at the bottom, bubbles of steam suddenly form, which radically changes the overall balance by redistributing the heat: ebullition accelerates and disrupts the prevailing order considerably, and in a nonhomogeneous manner. This phenomenon and others like it—so-called cavitation—can prove to be quite violent and sometimes inflict severe damage (on turbines, for example).

An hourglass will serve as our second example. Everyone has observed how flow regimes formed by the sand falling down the cone can vary. If, at the outset, the cone takes shape with a regular flow of sand at its periphery, all it takes is a slightly sharper angle—or a different kind of sand—for the flow to yield intermittent mini-avalanches, which drastically change the picture. The same principle is at work in a fusion plasma. Beyond the microscopic phenomena (turbulent or not) that govern the flow of heat or particles, more fitful phenomena can occur on spatial and temporal scales that are generally more macroscopic. Thus, the turbulent transport described above ("mini-avalanches") is capable of producing a negative effect on confinement as a whole.

On a more structural level, the configuration of the "magnetic box" in a nuclear device can become unstable either in part or entirely, quickly losing a significant fraction of the energy stored in the plasma. The branch of physics that studies the macrostability of magnetized plasmas is called *magnetohydrodynamics* (or MHD). It focuses on analyzing and predicting the stability of overall equilibrium in a hot and dense plasma confined along magnetic field lines and traversed by a current. The kinds of instability that have an impact on the plasma are many, and their effects vary in severity.

Much more could be said on the subject, but doing so would lead too far afield. It's vital to guard against these two major forms of instability in order to maintain plasma fusion in a tokamak.

Disruptions

Detractors of magnetic fusion often point to the first difficulty, which concerns disruption in the magnetic configuration. At issue is an extremely rapid magnetohydrodynamic phenomenon to be observed in tokamaks when, in a few tenths of thousandths of a second, the confinement goes missing and the totality of the kinetic and magnetic energy of the plasma "drains" into the surrounding structure. Given the speed at which this occurs, the forces and powers at play are very high and can threaten the material state of equipment.

In practical terms, the enclosure and its structural supports experience an intense surge tied to the loss of energy confinement. In turn, this energy occupies the walls in the form of transitory but consistent fluxes of heat, which are localized in keeping with how the disruption unfolds. Next to appear are forces tied to the abrupt disappearance of the plasma current, even though the toroidal magnetic field remains present. This magnetic energy will dissipate in the form of radiation and currents induced in the structure. In this phase, the plasma has abruptly cooled and its resistivity has suddenly increased; hereby, one observes a significant rise of the parallel electrical field that ordinarily underlies the inductive plasma current. Such conditions can provoke the intensive acceleration of an electron beam, which detaches from the rest of the plasma in a short span of time and terminates its course somewhere on the internal components of the enclosure. These three cascading phenomena represent a potential danger to the durability of the tokamak's vacuum chamber. Likewise, tokamaks regularly display damage due to disruptions ranging from physical impact on plasma-facing materials to changes in the internal structure from electromagnetic forces; the latter are capable of warping certain parts significantly. Other problems include local overheating deeper in the device, especially at the level of coils in the toroidal magnetic field.

Consequently, one must avoid situations that lead a plasma to experience disruption, or, when it's inevitable, detect the signs of disruption in advance in order to slow it down. At present, the research community is devoting intensive efforts to the matter. The task is to document, through experimentation, and catalog the conditions under which disruptions occur. Doing so yields a veritable map of the stable and unstable operational domains in the tokamak, which in turn facilitates both steering and modelization. For a few years now, this body of information has been greatly enriched by more and more specific information on detecting the so-called predisruptive phase, thanks to improved spatial and temporal resolution in diagnostic procedures and rapid data collection. Even though disruption occurs very quickly and can stem from multiple sources, it always announces itself beforehand; more and more, scientists are detecting and interpreting these signs in real time. Equipped with this information—even if it is not (yet) known how to adjust control parameters to counteract an incipient disruption—scientists can engage the buildup of a disruption in such a way that the most harmful effects will be avoided.

Several methods for mitigating disruptions are now being investigated. One of the most promising approaches involves smothering the predisruptive plasma by means of a massive and rapid jet of gas, which increases the temporal constants of the disruption and makes it easier to control. Drawing an analogy to aeronautics, researchers speaks of a "soft landing" in this context. Needless to say, experimentation is being complemented by concerted efforts to devise models. Recent advances in supercomputers have shed light on the nonlinear magnetohydrodynamic process at work in disruptive phenomena, which helps us to understand the brutal rupture of confinement that takes place, and to combine it with other techniques such as injecting gas into the plasma or using electron beams to control the sequence of disruption and mitigate its effects. Further work is still necessary, but efforts have opened the prospect of steady and reliable progress.

For engineers, the stakes are just as clear: the internal structure of tokamaks has to be made as resilient as possible to disruptions, and this is a matter of conception and design. From the outset, tokamaks are meant to endure a high level of induced forces; anticipating the currents induced in mechanical structures when disruption occurs represents the most crucial consideration for minimizing attendant constraints. Pertinent data from a wide array of sources is available, thanks to major advances that recent software and hardware have enabled in the mechanical modeling of complex structures. In addition, protecting plasma-facing components represents a more delicate undertaking. Although we are able to predict the fluxes of power attending normal operation and therefore to model the plasma's immediate periphery, the constraints arising from disruption that affects material components exceed what the latter can handle without damage. In this case, no real technological interventions exist yet to prevent a runaway disruption.

Does this threat to the tokamak's structural integrity represent a safety concern? It's clear that fusion reactors, because they will create and consume tritium, will de facto belong to the nuclear sector and therefore be subject to strict regulation. The matter concerns the physical confinement of tritium—whether any amount, however tiny, can be released into the atmosphere. The question is certainly valid; researchers are now conducting studies and working to perfect reliable techniques for detecting, in real time, problems of disruptions in tokamaks. The viability of nuclear fusion in installations based on the tokamak principle represents an item of vital concern for the high-level mission of the ITER project. This particular tokamak is now being built at Cadarache, on behalf of an international consortium of thirty-four countries. The initiative will be discussed at length in the next chapter, especially its nuclear dimension and the objective of producing energy from fusion on an industrial scale.

To close our discussion of disruptions, we should remember that the research community is making substantial efforts to document (as it has done for the tokamak) the "alternative" path represented by the stellarator configuration. It is doing so largely because of disruptions. A priori, the stellarator can create the same kind of magnetic bottle as the tokamak, but without relying on a plasma current; disruptions are not an issue for this device. Research to remedy the Achilles' heel presented by tokamak technology is occurring in tandem with the ITER project; the outcome will represent an important option for designing future reactors.

ELMs

The second magnetohydrodynamic instability in tokamaks that has to be fully understood and controlled before a reactor can be built is known by the barbarous name of *edge localized modes*, or ELMs. Above, we noted that transport-heat barriers such as the H-mode improve energy confinement. The phenomenon of ELMs is tied to this same barrier, which emerges at the plasma edge beyond a certain level of power. The pressure that builds up at the periphery of the plasma presents an extremely strong gradient, with the pressure profile resting on a "pedestal." We've also noted that under these conditions (which produce an effect similar to what occurs in an hourglass) the steep profile naturally leads to part of the plasma pressure being released, in order to return the system to a less unbalanced state. Indeed, at regular intervals, the plasma develops "puffs" of heat and particles toward the walls of the enclosure. These puffs are

ELMs. To draw an analogy (which shouldn't be taken too far), they represent what solar eruptions are to the sun, but for tokamaks.

What's the problem, then? Why should we worry about, or try to counteract, this occurrence? For one, an ELM is capable, on its own, of releasing a certain percentage of the kinetic energy stored in the plasma ring. The phenomenon isn't much in comparison to a disruption, but all the same—and just like a disruption—it's very abrupt and tends to repeat several times per second. Even though there's no need to worry about the same levels of structural stress, the situation is still serious inasmuch as plasma-facing components are subject to intermittent fluxes of great heat. In fact, ELMs generate a major—and cyclical—quantity of heat and particles, with fluxes that surpass what materials can withstand, at least over the long term. If nothing is done to weaken or avoid them, ELMs will erode components exposed to the plasma and increase the need for maintenance operations.

As with disruptions, it's important to understand the phenomenon and find ways to mitigate (if not eliminate) it. The last decade has witnessed many experimental studies on modelization, which have revealed key aspects of the instability at issue. First and foremost—and again, as in the case of disruption—ELMs depend on the plasma current; a priori, they do not occur in the stellarator configuration. Second, it's possible to maintain a tokamak plasma without ELMs within a certain range of plasma parameters. Finally, ELMs are sensitive to any number of disturbances, even tiny ones, that affect conditions at the plasma edge. Several forms of mitigation are already available for tokamak operations. Depending on the design, the size and/or frequency of ELMs can be reduced by changing the magnetic configuration at the outermost plasma edge. This is done by installing current coils at the periphery of the plasma. It's also possible to trigger an ELM by artificial means before it goes off on its own, at a point where the level of energy released will be acceptable. Just as scientists have developed means for countering disruptions, they have created prototypical solutions for dealing with ELMs, which are employed on tokamaks in operation across the globe; the goal is to implement the most advantageous and efficient midterm methods on the ITER tokamak. Sustained and coordinated modeling efforts are underway internationally, aiming to optimize our understanding of the finest details of the physics underlying these phenomena.

The upshot of our brief foray into the world of macrostability—especially in tokamaks—is that the physical parameters in which a plasma can achieve fusion under stable conditions hardly open onto an infinite horizon of possibilities. Once limitations have been determined, operating a tokamak in a stable realm requires particular tools, information that can be interpreted in real time by the piloting system, and the capacity (also in real time) to modify feedback on the plasma within predefined operational limits. A matter of such importance justifies the last two or three decades of experimentation, modelization, and development, and it illustrates the scope of the immense scientific and technological quest now underway.

Maintaining Performance over the "Long" Term

Once the plasma is preformed in terms of density, current, temperature, and other parameters enabling it to reach the conditions necessary for triggering fusion reactions among its components, a problem surfaces: maintaining it in this state. Our goal is to generate electricity, after all. We need to know not just how to provoke fusion reactions but also how to keep them going for a duration appropriate to the plasma's size—days, months, or even years.

One approach to the challenge posed by magnetic fusion is to increase, bit by bit, the characteristic timescale governing each of the phenomena at play. De facto, we have already addressed the first three steps on this "ladder" in our discussion of the trajectory of particles, magnetohydrodynamic stability, and heat confinement. In more quantitative terms—given their mean energy—electrons basically move at the speed of light. This means that they typically complete one cycle around a magnetic field line in one hundredth of a billionth of one second; in a reactor, they finish a cycle in about one billionth of one second.

The second step on the timescale is where magnetohydrodynamic instabilities appear—disruptions or ELMs. The characteristic time here ranges from a few thousandths of a second to a few dozen thousandths of a second—one tenth of the time it takes to blink your eye. Measurements, interpretation, and the feedback of actuators must be adjusted to this cadence in order to keep the plasma in a zone of stability. Clearly, the quest for continuous operation of a fusion plasma involves maintaining it in a favorable situation, like a tightrope walker keeping his or her balance. This difficulty has a positive side, too. If the plasma doesn't stay in a stable state, the reaction can only have one outcome: shutdown. A fusion plasma can never take off on its own or become uncontrollable—in contrast to bombs or the cores of fission reactors. In other words, the fusion reactors of the future will be inherently safe; needless to say, this represents a major point in their favor.

The third characteristic timescale concerns energy confinement, which is what enables phenomena of turbulence to transport heat from the core to the edge of the plasma. In the largest tokamak now in existence, the JET, the confinement time is on the order of one second. The next generation of tokamaks, whether as part of ITER or installed in future reactors, will have values on the order of five to ten seconds, at most. Once this limit is exceeded, the plasma's performance levels have been established; provided that Lawson's criterion is met, fusion reactions now are taking place. That said, the plasma has not yet achieved equilibrium; this is because of other phenomena that evolve more slowly, to which we now turn.

Although we have discussed energy confinement at length, we haven't addressed the vital matter of particle confinement. The two forms of confinement are not the same. Although particles carry the plasma's thermal energy on a microscopic level, the combined mechanism of individual trajectories (which are closely tied to magnetic field lines), collisions, and turbulence in the medium makes particles move from the core to the edge of the plasma (or vice versa) more slowly than the heat travels. Once again, the relation between the two timescales depends on multiple factors, but it typically lies on the order of ten. Particle confinement plays a key role in determining the maximum length of time for which a fusion plasma can maintain itself, and for two reasons. First, as reactions occur at the plasma core, the "fuel" (deuterium and tritium) dwindles, while "ashes" of helium-4 build up and dilute the mixture of reagents. If no steps are taken—or, rather, if particle transport proceeds too slowly—the outcome is fatal: the reaction goes out, smothered, as it were, by its own ashes. Clearly, a compromise must be reached between energy confinement time and particle confinement time so the plasma can constantly renew itself with "fresh fuel" while getting rid of "ashes." The trick is to find situations in which the energy confinement time lasts as long as possible without the particle confinement time growing too large.

In this context, we should look at how the density and composition of plasma are controlled in a tokamak. There are several ways to "feed" a plasma. As we have seen, one method is to supply additional heat by injecting fast neutral atoms. In effect, doing so makes it possible for deuterium and tritium nuclei with a great deal of energy to penetrate deep into the plasma core while heating the plasma as a whole. That said, under the conditions that typically prevail—which, ultimately, represent our real point of interest—the number of fast atoms delivered per second will fall far short of the level at which deuterium and tritium nuclei are consumed by the fusion reaction itself. Increasing the flow of fast neutral atoms would mean increasing the power added or decreasing the individual energy of atoms—in other words, adding further constraints. Other methods for feeding plasma are necessary, then, and with neutral atoms; only they are capable of thwarting the barrier of magnetic confinement from outside.

To this end, there are two tried-and-true techniques, which now qualify as "classic." The first involves jets of neutral gases at varying intensities applied to the periphery of the plasma. The second method is to inject deuterium or tritium "ice cubes" at extremely high speeds into the periphery, in the manner of a blowgun. As futuristic as all this might seem, these technologies have been progressing by leaps and bounds for more than two decades. Deuterium or tritium gas can be cooled to the point of forming a solid, at a temperature below –259°C. A few cubic millimeters in size, the "ice cubes" contain enough matter to feed the present generation of tokamaks when they are injected at a rate of ten per second, at speeds of several hundred meters per second. This method, which is extremely reliable, is in the course of being adapted for future generations of machines. These tokamaks, bigger and hotter than their counterparts today, will require even faster injections for the ice cubes to reach the core. Size is a less important parameter when offset either by increased frequency or by a greater number of injectors.

The second challenge researchers face when examining the balance between what goes in and what goes out of equipment is the question of evacuating particles at the periphery of the plasma. Picture the torus of hot plasma, surrounded by the confinement unit. Plasma that has been partially consumed—composed of deuterium, tritium, and a certain amount of helium-4 "ashes"—is very close to the enclosure at its outermost points. Here, it's relatively dense and "cold" compared to the core (tens of thousands of degrees, as opposed to hundreds of millions of degrees). If the plasma edge is "skimmed" and a portion removed at the same rate that fresh deuterium or tritium is injected, it will be enough to keep the mixture sufficiently low in helium ashes for reactions to continue. This is the operative principle for all fusion machines—both tokamaks and stellarators.

Once the balance between feeding and draining the plasma has been resolved, thermal equilibrium between the plasma ring and its immediate surroundings (the vacuum chamber) must be ensured. Although the magnetic bottle lets one confine the plasma's energy in order to raise the temperature and facilitate fusion, it also slows down the transfer of heat from the core, where it is created, to the periphery. Confined plasma is not insulated from its environment to the point where no thermal exchange takes place. Quite the opposite. In the reactors of the future, the heat created by fusion will have to be recuperated at the end of the line by steam generators in order to make turbines revolve and produce electricity.

At present, the question of how to manage fluxes of heat at the periphery of equipment is the major issue confronting the international research community. For a sense of the scope, here's a number: ten million to twenty million watts per square meter. That's how much heat is continuously transferred from the core to adjacent material components. In effect, if part of the energy produced by fusion reactions ends up as neutrons and electromagnetic radiation—which both exit in isotropic manner and strike the walls at the level of a few million watts per square meter—the rest of the energy conducted by the plasma concentrates at very specific points and reaches a level between ten million and twenty million watts per square meter. To get an idea of what this means, the heat flow on the surface of a star like our sun is "only" two or three orders of magnitude higher than this value. Putting hardware components in front of a tokamak plasma amounts to asking engineers to devise a spacecraft that can land on the sun—and, what's more, stay there indefinitely! Both aspects of the problem count. Everyone knows what happens when you put your finger close to a candle; the finger only gets burned if it stays near the flame for a certain amount of time. In scientific terms, the challenge involves the energy density to which plasma-facing equipment is subjected, that is, the product of the power density and exposure time.

The situation may be summed up as follows. Even though the pioneers of fusion quickly recognized the problem, it didn't bother them much for the simple reason that the life span of the first plasmas—a few thousandths of a second—was so low that the energy deposited on structures could be dealt with easily. Subsequently, tokamaks developed the capacity to generate and maintain fusion plasmas for longer durations: a few seconds in the 1980s and then a few minutes. Only now did the technological problem of equilibrium between the plasma and its immediate physical surroundings (that is, components facing the plasma) emerge.

Two distinct issues had to be addressed and optimized concurrently. The first concerned managing the flow of impurities generated when plasma ions bombard material surfaces. The phenomenon is fairly complex, since it involves both plasma ions sticking on the structure (so-called retention) and extraction of atoms from exposed surfaces (which is known as *erosion*).

Whereas a deuterium or tritium atom entering the plasma brings along one electron per nucleus, an iron atom (for instance) will bring with it twenty-six electrons and then be virtually "substituted" for twenty-six deuterium or tritium nuclei. This dilution of the mixture of reagents by "heavy" impurities will affect the plasma's reactivity if nothing is done, and the fusion performance will diminish correspondingly. What's more, atoms as heavy as iron—or other metals, such as tungsten—will not necessarily ionize completely in the plasma, since atoms occupying less peripheral positions are generally tied to the nucleus with energy levels greater than the local temperature of the plasma. Such heavy, partially ionized atoms retain the ability to transform part of the energy stored in the plasma into radiation through processes of excitation and disexcitation within this residual electron cloud. As a result, by becoming charged with impurities, the plasma will start to radiate part of the energy in its core, which directly lowers the energy confinement time and weakens the fusion performance level.

To compensate for the imbalance that occurs when the deuterium-tritium mixture is polluted by heavier ions from the tokamak's walls, efforts have been made to cover surfaces exposed to the plasma with materials that are highly refractive and have a very small atomic number. Carbon quickly came to represent the standard: with only six electrons and remarkable physical properties, it is excellent for refracting heat and doesn't melt. Since the 1980s, the inner walls of more or less all tokamaks in operation have been covered with graphite components, which makes it possible to contain plasmas that are less vulnerable to heavy, radioactive impurities and maximize performance correspondingly. The 1990s witnessed the emergence of plasmas with durations often surpassing ten seconds; this development called for modifying graphite components and actively employing cooling elements. The same period saw the birth of several generations of technology, up to the CEA Tore Supra. This machine incorporates carbon-fiber tiles enclosing a matrix of graphite, welded to a copper cooling channel. The unit components are arranged to form surfaces capable of withstanding heat flows being completely stationary. Heat from the plasma that reaches these surfaces is immediately conducted toward the cooling channel by the carbon fibers. Hereby, plasma-facing components do not reach temperatures beyond a few hundred to a thousand degrees—a level that poses no danger to them.

While universally welcomed, this accomplishment harbored two shortcomings that impeded further progress. As noted above, carbon exposed to plasma will erode. Even if pollution of plasma remained low and at a level acceptable for ongoing fusion, the speed at which components would degrade in a reactor built along these lines would pose maintenance problems, affecting its functional availability. What's more, repeated tests on tokamaks like Tore Supra have shown that—because of the chemical affinity between carbon and hydrogen isotopes (deuterium and tritium)—the process of erosion and the deposits it entails lead hydrocarbonized compounds to accumulate on the walls of equipment. Although this phenomenon has not yet been fully documented, effort should be made to avoid it in future reactors. As a potentially significant cause for tritium getting trapped on the walls of the tokamak, it represents an undesirable factor.

For these reasons, the last ten years have witnessed decreased use of carbon at the periphery of fusion plasmas. With an eye toward constructing ITER, the European fusion program has launched a comprehensive scientific and technological program to test the only alternative to carbon that the periodic table offers: tungsten, which combines refractive properties with a very low rate of erosion and retention; this is the element used to make the filaments of incandescent lights.

Tungsten research affords a further example of the remarkable spirit of collaboration that prevails in the fusion community. At the dawn of the millennium, scientists recognized that the carbon solution they had been working on for some two decades would lead to a major impasse when building a reactor. ITER is already well underway, and both its scientific operating point and its practical success will depend on choices made now. The level of plasma performance anticipated represents the fruit of extrapolations based on energy confinement in tokamaks currently in operation; such confinement depends on the particulars of how plasma density and pressure are influenced by conditions in the immediate periphery.

Observation of carbon plasma-facing components led researchers to envisage an operating point for plasma with relatively low edge densities and relatively high edge temperatures. But using a metallic compound for walls required that plasma edge conditions reach higher densities and lower temperatures. German researchers were the first to rise to the challenge by covering the carbon components of the ASDEX-Upgrade (ASDEX-U) tokamak with thin layers of tungsten. This inexpensive and quick measure produced favorable results. Likewise, scientists at JET—which is still the largest tokamak in operation and the most important site for extrapolating operating points—completely rebuilt plasma-facing components by integrating tungsten in a geometrical design close to that of the future ITER. After a decade of testing, they were able to reposition the operating point expected of the ITER at a level close to what had been anticipated, thereby confirming the plasma's compatibility with a tungsten environment. That said, all these experiments were conducted in nonstationary manner on just two tokamaks, ASDEX-U and JET, neither of which was equipped to support long-duration plasmas. As such, it wasn't possible to verify whether the tungsten components themselves were compatible with plasma—that is, whether they would hold up under other conditions.

Thus, since 2012, the French fusion program has been modifying the Tore Supra tokamak (a machine that can create plasmas for minutes at a time, and do so repeatedly) in order to obtain the conditions necessary and sufficient for testing prototypes of tungsten plasma-facing components that are actively cooled. As this book goes to press, experiments are still being conducted; the results will prove vital for manufacturing tungsten equipment for ITER. In tandem with technological advances, scientific progress has been made, notably by integrating details about plasmas and materials into computerized models. Even if a long road lies ahead, the goal has been clearly identified and a significant portion of the global community is hard at work.

Resolving the problem of thermal equilibrium in the plasma and its surroundings both technologically and operationally will open the door to ITER—the object of so much political and scientific attention in our day, and the subject of the next chapter.

7
The Star-makers

The ITER project marks the beginning of the third major phase in the development of fusion energy: demonstrating its feasibility on the scale of a reactor and showing the fullest degree of control over relevant processes.

To draw up a very quick balance of advances in magnetic fusion research until this point, we can say that now, at the dawn of the twenty-first century, we have obtained the basic intellectual capital needed to define the physical parameters for triggering fusion reactions here on earth and controlling more or less all the techniques and technologies involved. ITER represents, for the first time on record, the site where the ensemble of elements can be brought together and assembled on the actual scale required. The task, then, is to confine a plasma composed of a mixture of deuterium and tritium in a tokamak of sufficient size and sophistication to reach—and maintain for periods on the order of an hour—a net increase of power at the plasma core fixed at a target value of ten.

To appreciate the immense challenge at the heart of ITER, we should call to mind the best performance levels attained by tokamaks past and present.

First, in terms of the performance of hot plasmas, tokamaks such as TFTR, which was in operation in Princeton until the 1990s, the Japanese JT-60U, which closed down ten years ago, and JET, which is still up and running in England, have all reached core plasma temperatures on the order of one hundred million to one hundred fifty million degrees; such values are a priori sufficient. However, the plasma volume magnetized by these machines never exceeded one hundred cubic meters, and the plasma current never went beyond a few million amperes. In other words, they operated with a maximum confinement time on the order of one second—which approaches the conditions imposed by Lawson's criterion marginally, at best. At the end of the 1990s, JT-60U and JET created plasmas that came quite close to satisfying this criterion, but only JET and TFTR were outfitted for manipulating and injecting tritium experimentally. In consequence, JET is the sole tokamak to have "really" produced fusion power at the level it takes to heat the plasma.[1] This operating point is called *breakeven*. In 1995, JET generated sixteen megawatts of fusion power, which is still the world record. The target for ITER lies between four hundred and five hundred megawatts, employing ten times as much plasma with a projected tenfold gain in return. The power generated by fusion reactions will be thirty times as great—a considerable step forward.

In matters of technology—and also since the beginning of the 2000s—research and development that started decades ago has made it possible to envisage constructing such a giant. Tokamaks built in the 1980s that were devoted to studying plasma performance sought to maximize plasma parameters for only very short durations and therefore could not achieve, much less maintain, relevant performance levels for more than a few seconds. The problem lay with the technology employed (copper field coils, nonrefrigerated cooling, mechanisms for adding heat with short impulses, diagnostic systems unequipped to manage heavy neutron fluxes or significant heat, limited postprocessing of data). Equipment needed to be improved in keeping with scientific advances in order to arrive at the conditions necessary for the ITER project.

Above all, the tokamak required would have to generate a static magnetic field in continuous manner. The thousands of amperes circulating in toroidal and poloidal coils mean that conventional technologies cannot be extrapolated because of overheating due to the Joule effect. Even when standard wires are cooled with circulating water, losses affect the installation's overall yield and therefore call for other methods.

At the beginning of the 1980s, scientists got the idea of developing tokamak magnets by enlisting new, low-temperature superconductive technology. The size required for magnets was immense, and only a small number of teams took on the challenge. But in record time, CEA researchers rose to the occasion and developed the Tore Supra, a tokamak with toroidal coils made from conductive strands of niobium and titanium encased in a copper matrix, all of which was bathed in helium at 1.8 K. At this temperature, the alloy conducts an electric current without any resistance, and the helium requires no pump to circulate among the coils and flush away local sources of heat. For thirty years now, Tore Supra has used this magnet, demonstrating the system's feasibility and showing the scientific community how to move forward. This technological accomplishment is what opened, in the 1990s, the technological path to ITER, even if the solution ultimately adopted has changed somewhat in light of size constraints. Indeed, its success has produced notable effects beyond the realm of fusion. The techniques employed have also been used on the Large Hadron Collider at CERN (European Organization for Nuclear Research [*Conseil Européen pour la Recherche Nucléaire*], founded in 1954) and for Iseult, the high-resolution medical resonance imaging system being set up at NeuroSpin (Plateau de Saclay, France).

Once the first obstacle to a permanent magnetic field was resolved, scientists had to ensure that means for sustaining the plasma current—and therefore the poloidal magnetic field generated by the plasma itself—would be operational for long stretches of time, if not continuously. The numerous technological developments made in this context represent the fruit of close cooperation between an extremely rigorous research community and a global industry attuned to challenges and innovations in the field. Over some twenty years, microwave tubes individually capable of generating sustained power on the order of one megawatt (tetrodes, klystrons, and other gyrotrons) have been developed; their frequencies now span more than four orders of magnitude. Likewise, techniques for creating and accelerating negative ions have made spectacular progress, opening the way for systems that inject fast neutral atoms at levels up to one million electron volts. Finally, methods for managing the evacuation of power flows that affect plasma-facing components and coupling the additional power sources discussed above have put the pieces in place for completing the technological puzzle.

With these elements already up and running or in the course of development in the 1980s—when JET and Tore Supra were first being put into operation—fusion scientists began their march toward the next, crucial phase, which would soon yield the ITER project.

Every now and then, felicitous encounters occur between History, writ large, and endeavors on a smaller scale. This was certainly the case for ITER, which benefited from the good graces of two powerful sponsors: Ronald Reagan and Mikhail Gorbachev. The story begins at the Geneva Summit held in November 1985 when the two heads of state met for the first time and set the process into motion that would lead to the end of the Cold War. On this occasion, numerous fields of investigation were opened that confirmed the role of scientific cooperation in building alliances and making peace. ITER was ideal for symbolizing the shared goals of two superpowers. Exchanges born of this momentous political reconciliation soon intensified, including first France and the United Kingdom and then Japan. The birth certificate, as it were, appeared in documents of the European Commission, signed April 21, 1988. The project's contours were still vague, but history was in the making. Project development fell to the International Atomic Energy Agency, and the first team members set to work at the Max Planck Institute for Plasma Physics at Garching.

ITER made its debut, then, as a scientific and technological project of international dimensions. The technical team's first objective, with fusion laboratories elsewhere lending support, was to draw up preliminary plans for mapping out what needed to be done and the resources it would take for the tokamak to be built. ITER issued its first major report in 1992, a detailed account of steps to be performed over six years, at a cost of one billion dollars; expenses would be covered by a consortium joining the United States, the Soviet Union (which would soon become Russia), Japan, and the European Union. Further construction expenses were anticipated in the range of five to six billion dollars. In the money of the day, that's not too far from the final consolidated costs. Paul-Henri Rebut, who had designed and constructed the TFR and JET tokamaks, was put in charge of the project. Accordingly, the initiative benefited from collaboration between three sites: Garching, San Diego, and Naka. Ever since, the sun—the symbol and goal of this unprecedented undertaking—has never set on ITER!

Before long, however, difficulties emerged in two areas, in keeping with the originality and uniqueness of the situation. The first set of troubles was internal in nature and foreshadowed what would prove to be the Achilles' heel of the project. For an enterprise of such scope, roles and responsibilities have to be defined as explicitly as possible; failing such measures, conflicts arise apropos of decisions recommended by the project team, on the one hand, and the financial considerations of funding partners, on the other. Born of political will, ITER was an international research project without organizational precedent. In this light, we can see how the partners didn't really want to grant its director full discretionary power over technical and financial decisions. It was constantly necessary to strike a balance between technical needs and political and monetary demands. In consequence, a series of crises shook the project, especially when technical choices gave partners reason to fear that costs would prove to be higher than expected. In 1994, Rebut stepped down. Robert Aymar, who had designed and built Tore Supra, took his place—with the provision that better cost control would guide the facility's conception. By the end of 1997, a plan was in place, and attention could turn to the question of construction.

With some vehemence, difficulties in the second area then arose: external problems. We should bear in mind that the timescales of fusion don't automatically coincide with the pace of political events in a fast-moving world. ITER may have benefited from improved relations between East and West in 1985 and from the ongoing effects of a worldwide petroleum crisis, but the end of the 1990s witnessed an unusual upturn in the petroleum market and societal wariness of nuclear energy—especially in the wake of the Chernobyl disaster. The first defection, which represented a veritable bombshell, was on the part of the United States, which announced in 1998 that it was withdrawing because of issues with the distribution of research funds by the Department of Energy; the ITER project was competing with national initiatives, especially the National Ignition Facility, where researchers were investigating inertial confinement fusion. Even as the details of ITER were being worked out technically, negotiations about the site of operations and financing its construction were dragging and threatening the project's very existence.

The remaining partners stayed the course and focused on potential sites at Garching and Naka. At the same time, the project director was instructed to review the scale of facilities with an eye to reducing building costs. In 2001, revised plans were ready—albeit with a machine that would be smaller than the one initially envisioned. As we have seen, size and performance levels are related for tokamaks; the new design foresaw lower performance margins, with the objective of amplifying power by ten instead of thirty (as originally planned). By the same token, it became increasingly clear that the goal of generating self-sufficient amounts of tritium would have to be abandoned. As part of the ITER program, partners would need to develop prototypes of, and test out, different means of fabricating tritium on-site by bombarding lithium targets with fusion neutrons; a tokamak capable of generating and using its own tritium in a closed cycle was put off until the next stage (which we will discuss in the final chapter).

It was only in mid-2001, then, sixteen years after the meeting between Reagan and Gorbachev, that partners—Europe, Japan, and Russia—started looking for a construction site, all the while making efforts to expand the ranks of countries involved. The overture proved fruitful. Somewhat surprisingly, the first solid offer came from Canada. Here, a private industrial consortium proposed building ITER at Clarington, near Toronto. The proximity of CANDU (Canada Deuterium Uranium) fission reactors producing tritium represented a point of additional appeal, since it was now clear that ITER would not have a sufficient supply of its own. Although the offer didn't enjoy the support of the Canadian government and, as such, had little chance to win over ITER members, it catalyzed a more active search for other locations.

While Japan was preparing to propose a site at Rokkasho (Aomori Prefecture), Europe insisted on either Cadarache (in the vicinity of the Tore Supra tokamak operated by the French Atomic Energy Commission) or Vandellòs, Spain, where a nuclear plant already existed (near the city of Tarragona, Catalonia).

Several books have recounted the quasi-Homeric events that ensued. They put into relief the remarkable spell that the ITER project cast over scientists charged with bringing efforts to fruition, the populations of the countries involved, and political figures—all in the name of an adventure that would transcend the standing divisions between the world's inhabitants. On November 26, 2003, it was finally announced that Cadarache would be the European standard-bearer, while Spain would host the legal entity responsible for constructing ITER (the so-called Domestic European Agency, later christened "Fusion for Energy," which has had its seat in Barcelona since 2007).

Pleasant surprises occurred in 2003 in terms of sponsorship. Five years after its shocking withdrawal, the United States announced that it was back on board in January—just after China formally requested to join the project. Six months later, when South Korea became a member, the new ITER roundtable included countries representing about half the world's population and commanding 80 percent of gross domestic products across the globe. Things were moving again.

Rivalry between advocates of the proposed European site of operations at Cadarache and proponents of its Japanese counterpart at Rokkasho proved bitter, however. It didn't take long for two blocs to emerge: Europe-Russia-China, on the one hand, and Japan-South Korea-United States on the other. Countless meetings were held prefiguring the ITER Council that would finally emerge: bilateral negotiations and repeated visits between participants often working at cross-purposes. The diplomatic ballet lasted for some two years, and 2005 began with little hope for a resolution. The competition between Europe and Japan grew quite severe, with mounting tension and ultimatums on both sides; it seemed ITER might never get built.

Then, in March 2005, Jacques Chirac made a state visit to Japan, bringing along the project's all-important dossier. It's hard not to admire the skill of the president of France, who demonstrated considerable knowledge of Japan and its culture. Chirac's gambit worked. Immediately afterward, the situation de-escalated and negotiations proceeded with a promise of "win-win." A few months later, Europe and Japan finalized their agreement to make Cadarache the site for ITER and to adopt an expanded approach that would strengthen collaboration by adding three (cofinanced) projects on Japanese soil. The JT-60U tokamak, which was not operating at the time, would be updated into a superconducting tokamak baptized "JT-60SA." Partners would jointly develop a prototype for irradiating neutron-based fusion materials, IFMIF-EVEDA (International Fusion Materials Irradiation Facility Engineering Validation and Engineering Design Activities). Finally, they would found the International Fusion Energy Research Centre, a fusion research center located at Rokkasho, where planning for future reactors and high-performance computer simulations of plasmas would take place.

On June 28, 2005, a vote at the ministerial level of the ITER Council ratified the decision to construct ITER facilities at Cadarache. The occasion was almost twenty years to the day after Reagan and Gorbachev's historic meeting.

India joined the ranks of participating nations on December 6, 2005. With this seventh megapartner, more than half the global population now supported ITER, a vital symbol of international cooperation as well as a vector of innovation and dialogue between the peoples of the world. At this juncture, a new director general, Kaname Ikeda, was also appointed.

The ultimate success of the French site calls for a few words of appreciation for those who shepherded its candidacy, notably the CEA[2] teams in charge of researching the location and political leaders such as Claudie Haigneré, minister delegate for Research and New Technologies from 2002 to 2004, then minister delegate for Euorpean Affairs from 2004 to 2005.

On November 21, 2006, the agreement officially creating ITER was signed at Elysée Palace by ministers representing the seven partners. The meeting was convened by Jacques Chirac, the project's most prestigious sponsor, under the auspices of the International Agency of Atomic Energy, which thenceforth held custody of the document.

The following year, the ITER International Organization, charged with implementing and managing operations, was formally founded. Agencies in partner states ratified the treaty and named Kaname Ikeda director general. The first paid employees now moved into provisional facilities at Cadarache, as France saw to constructing the building that would be the administrative center. Coordinating efforts internationally, the various partners created domestic agencies to provide equipment and services, as per the organization's charter.

The initial division of responsibilities and tasks was fairly complex. To appreciate the issues that arose, we should bear in mind that the detailed designs completed in 2001 had projected the overall costs. The negotiations that followed, about participation and the site of ITER, concerned how expenses would be distributed among members. The originality of the project (without which nothing would have happened), but also its most complex aspect, was that partners insisted that payment be provided "in kind" by the various domestic agencies, and not in cash. In other words, given the high costs, the technological and industrial complexity of the undertaking, and the long-term economic stakes of a new energy source, member states all wanted to work on developing key elements themselves. Consequently, the project was "carved" into parts that each partner would have to provide, in keeping with a complicated system for balancing out competing interests with overall expenses. Well before Bitcoin, then, a virtual currency called IUA (ITER Unit of Account) was created, which, ever since, has served for transactions between ITER partners. The organization assigned a credit value in IUA to each subproject and distributed parcels in such a way that Europe, the host partner, would be responsible for 45 percent of the total cost, and the six other members for the remaining 55 percent in equal shares. The endowment provided for the international, executive branch of ITER covered only internal operations.

A great deal has been written on the approach taken, a modus operandi that stands at the source of many of the ills that befell ITER in its early years. Before going into the particulars, we should recognize the challenge, both diplomatic and technical, that creating the international organization entailed—especially in light of members' goal of coming away from the project with as much know-how as possible in their own possession. The backdrop for this great scientific enterprise was—and is—formed by the immense societal stakes of pioneering a new energy sector. In other words, the bumpier part of the ITER road is part of the price to be paid for the project going into fulfillment at all.

With this caveat in mind, we now turn to concrete aspects of the project—what happened when initial enthusiasm met up with reality. The first point of difficulty concerned the timeline. Plans had foreseen some ten years for construction. This projection failed to anticipate how long it would take to start from scratch and not just put together an international team of highly specialized experts working at the same location but also set up seven domestic agencies functioning at the same level in seven different countries. Even leaving aside administrative, juridical, and logistical aspects of the undertaking, no allowances had been made for people to "catch their breath." In fact, collaborators from across the globe needed to come to Cadarache and form a viable community capable of working together while, at the same time, belonging to a network of other institutional structures, which likewise would have to operate efficiently.

This difficulty was compounded by another problem, which was more technical in nature. Decisions—some of them a decade old by this point—had not been updated, yet now they were "written in stone." Needless to say, the research community had hardly been twiddling its thumbs in the interim. New operational scenarios and limits, to say nothing of technological innovations, had been developed, all of which had a bearing on the design and operation of the project. Consequently, the first head of ITER faced a slew of requests from various research laboratories to modify details of the plan; taken together, they led to major revisions. It isn't hard to picture the Pandora's box that resulted in terms of added costs and delays. Should everything go forward as outlined in 2001? Needless to say, it was impossible to change the past simply to suit present purposes, yet vital to consider long-term goals.

That said, even for situations this complex, means do exist for managing the problems that arise. Accordingly, ITER implemented numerous measures to describe and diagnose difficulties in the first five years after the project's launch.

By grappling with setbacks and unforeseen developments, both in terms of cost and delays, organs of control established by partners were able to analyze overall progress with more and more precision. Finally, having withstood small- and medium-sized troubles, ITER confronted a major crisis in 2014 when it was determined that plans were behind schedule and expenses mounting. According to figures, construction was costing twice as much as it should. Even when adjustments were made to account for differences between actual costs in 2014 and estimates made back in 2001, it was clear that the project was getting too expensive. Failure to adhere to the timeline also posed a real problem. In effect, facilities were scheduled to produce plasmas in 2020, even though the site still didn't have a single building for technical operations. Administrative inquiries and progress reports multiplied. The United States, in particular, indicated readiness to leave again if the situation failed to improve.

Against this backdrop, a new phase of ITER started at the beginning of 2015. Bernard Bigot was appointed head of the project, and a new plan of action with a number of forceful measures went into effect. The timetable was reworked from top to bottom, and the new director general received full discretionary power to act, over and above decisions made by domestic agencies in partner countries. Funds placed in his hands would cover any cost overruns that should arise. These changes produced spectacular effects, and construction went forward with renewed vigor. According to the revised timeline, buildings will be complete and facilities will go into operation at the end of 2025. Programs of experimentation (involving any number of interconnected subprojects) are intended to get everything to the working point in 2035. Construction is now in full swing, as is the manufacture of components by member countries. Delivery of technical equipment has already started via a special route between the harbor of Fos-sur-Mer and Cadarache, and several hundred special convoys are expected to be traveling down the roads of Provence soon.

ITER is a cathedral of technology. Once assembled, its tokamak will weigh twenty-three thousand tons. The superconductive magnetic field cage for confining the plasma will weigh ten thousand tons on its own; it

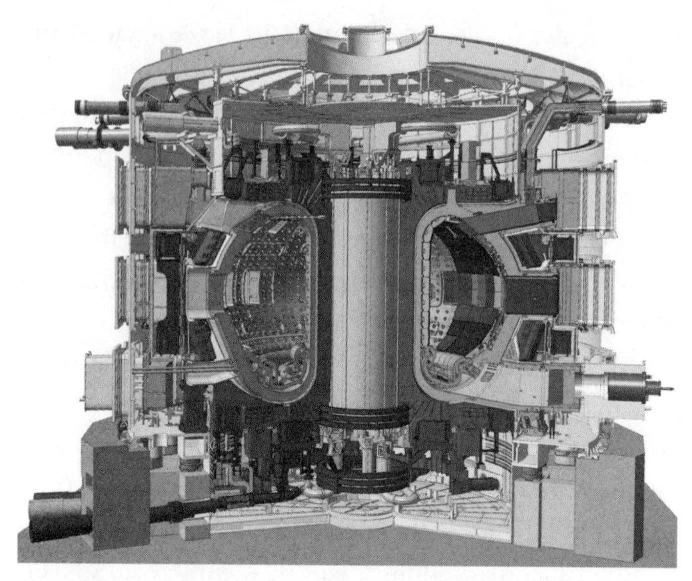

Figure 7.1
Exploded view of ITER (© ITER Organization).

will consist of eighteen toroidal coils seventeen meters in height, each weighing three hundred sixty tons, and six poloidal coils, of which the largest will be twenty-four meters in diameter with a central solenoid eighteen meters high. The toric vacuum chamber is designed to hold about one thousand cubic meters of plasma, which will be traversed by a current capable of reaching fifteen million amperes. Auxiliary heating systems for the plasma will be able to inject up to seventy-three million watts continuously at the initial stage, and up to one hundred ten million watts in the ultimate phases of the project (if necessary). Plasma-facing components will include a *far wall* cooled and covered by beryllium and a *close divertor* in a refrigerated block of tungsten for managing flows of heat and particles at the plasma edge. As mentioned above, ITER is intended to demonstrate, on a full scale, that it's possible to domesticate fusion reactions on the earth, generating between four hundred and five hundred fusion megawatts for durations on the order of ten minutes to an hour. Such levels admit comparison to what a nuclear fission reactor generates. In sum, plans hold for ITER to be supplied with tritium externally for daily operations. Partners will supply between four and six "tritigenic" components to cover the enclosure; these technologies prefigure equipment for tritium production in situ on future fusion reactors.

Research at ITER is supposed to go on for some twenty years and confirm (or disconfirm), both scientifically and technologically, the viability of fusion reactions on the scale of a reactor. If results are positive, they will open the way for industrializing the process. Like any nuclear installation located on French soil, ITER has to respect the rules established by the Nuclear Safety Authority (*Autorité de sûreté nucléaire*) in matters of design, construction, and operation. Accordingly, an international partnership has been set up to ensure the safety of the installation, which is unique in kind. Needless to say, measures include all aspects of maintenance while the site is up and running and foresee disassembly when the project comes to an end. It bears repeating: although fusion reactions of deuterium and tritium don't produce radioactive elements, tritium itself is radioactive; what's more, the neutrons that reactions generate activate materials they encounter and ultimately make the tokamak itself "nuclear waste." Therefore, ITER's mission includes demonstrating that the tritium management cycle can be controlled, and that it is possible to construct a tokamak using materials with a low activation rate under neutron irradiation, so that dismantling it will prove quick and easy, with a minimum of mid- to high-level waste. ITER's host country has taken charge of piloting these aspects of deconstruction—"getting back to nature" (*retour au vert*), as the saying goes. The attractiveness of the new energy source also depends on considerations of this nature.

8
. . . and Then?

> [. . .] one must imagine Sisyphus happy.
> —Albert Camus, *The Myth of Sisyphus*

Sixty years have passed since humankind embarked on the somewhat mad enterprise of domesticating the energy of the stars. Does this span represent a long time or just a little? Is it worth continuing to fund the enterprise? Does it have a reasonable chance of proving successful? Aren't there other solutions—easier ones, and less expensive too? Does the solution to the problem of energy even exist? Could this be another case of smoke and mirrors clouding the judgment of scientists?

All of these questions are legitimate. However, one shouldn't confuse motivations with solutions, especially where research is concerned. As has been stressed throughout the preceding, fusion research doesn't represent a fundamental discipline in the same sense as cosmology or high-energy physics. Instead, it has the concrete goal of serving social interest. In this regard, a moral contract holds between society at large and the scientific community, one that is made up of resources allocated by the taxpayer and global management independent of those seeking support.

That said, and as we have emphasized from the outset, fusion research joins the search for a new, clean, and practically inexhaustible source of energy with the Promethean undertaking of finding and mastering a "secret" with enormous implications for humanity as a whole. The concrete work of scientists and political decision makers anchors this socially and symbolically charged enterprise. Indeed, this is why the research community has managed to organize itself so well on an international scale from the very start, in contrast to any number of other fields with more "competition" and schedules of operation at smaller intervals. The almost familial spirit of cooperation that prevails is striking, as are levels of individual commitment when facing a superhuman task. Few areas of research display this twofold character: a quest for the Grail and a relay race in one. As we have seen, the task demands both scientific imagination and technological ingenuity; as such, it's a planetary enterprise that admits comparison to the conquest of space.

The ITER project does not concern knowledge alone, but also power. When it's finally clear that we understand the principal mechanisms governing plasma fusion and have mastered the relevant technology, what is to be done then? At this stage, the question of societal and economic stakes will be felt in full. Although it will take another two or three decades before we need to decide "what happens now," it's vital to start making preparations for the future. Even though plans won't enable day-to-day predictions, they will identify what needs to be done in shifting from ITER to commercial reactors, the amount of money and time involved, and what we can expect in terms of competition from other energy sources.

All ITER partners are busily drawing up "road maps," which use some of the same constants but incorporate major differences, as well. At the same time, positive results—with the prospect of more to come, soon—have sparked strong interest among private investors, who recognize that the energy market promises unusually high returns for quite a few decades to come.

Fusion as an Energy Source

Let's take a quick look at the "royal road" for fusion research—in other words, what needs to be done for the scientific results and technological gains expected from ITER to yield a deuterium and tritium fusion reactor operating by means of a tokamak.

Although ITER is supposed to provide us with the mode of operation and dimensional stability of plasma in fusion, we need to do at least two more things before building a reactor, even one that is still a prototype. The first is to control the full, closed cycle of tritium: its production on-site, its recuperation, and, finally, its reinjection into the plasma. Partner countries are in the process of developing different apparatuses and testing them at ITER facilities during the deuterium-tritium phase. Possibilities for industrial application will be determined in light of these tests. This factor has pushed final designs for the next stage up to the year 2035.

The second item absent from the ITER program is the material to be used for the reactor itself—for instance, the steel constituting the vacuum chamber. In order to be commercially viable, a tokamak reactor will need to operate with a high availability rate and have a life expectancy comparable to that of electrical plants already in existence (whether nuclear or not): at least forty years. In contrast to ITER, which has been conceived solely for research purposes and produces fusion neutrons only sporadically, the future reactor will operate for many years and generate a total quantity of neutrons several orders of magnitude higher; in so doing, it will stress structural materials at a level that those now in use cannot sustain. Needless to say, active programs of research and development are vital for conducting detailed metallurgical study and building experimental equipment that will enable scientists to subject samples to the neutron flows expected from a reactor. Several initiatives are already underway. They include the European-Japanese project IFMIF, which is developing prototypes of an accelerator that will propel deuterium nuclei at a target formed by a continuous film of liquid lithium; irradiating test materials with neutrons will allow their properties to be evaluated under extreme conditions. These "building blocks" (accelerator and target) are now in the validation phase. IFMIF members anticipate being able to assemble all components and begin experiments very soon, in tandem with ITER.

Such programs represent vital stepping-stones on the way to fusion, and ITER partners other than Europe and Japan are making contributions, as well. The highest scientific standards must be observed when researching and modeling materials, and just as much attention should be brought to bear on practical aspects. One alternative to the accelerator-target project involves a tokamak with dimensions close to the one at ITER but dedicated to producing a neutron flux in a systematic and repetitive manner; instead of being geared toward maximum performance, such a machine would operate with a much lower amplification factor than its ITER counterpart, in keeping with the task assigned. Any number of projects are on the books. Undoubtedly, the most serious one is the China Fusion Engineering Test Reactor project, which aims to provide not only essential information about irradiated materials but also systematic guidelines for maintaining fusion in future reactors.

ITER is already dealing with concrete maintenance issues—among other things, the intricacies of keeping complex nuclear components in shape. Over and above regulation, developments include remote-handling systems that employ robotic technology to assemble and disassemble components and conduct inspections. These measures were clearly motivated by research at JET toward the end of the 1990s, when operations with tritium, however infrequent, prompted a new understanding of the task at hand. For future reactors, which will be sharply constrained in terms of the energy available, maintenance factors must be developed and incorporated into designs from the outset in order to gauge overall impact. The most advanced study in this domain is now being conducted in Europe and China, and it has brought about major changes to the ways that future tokamaks will be designed and assembled.

Other steps are being taken to improve the tokamak configuration and make equipment more reliable. Management of heat and particle flow has been the object of intensive research, and some efforts are going back to the very "source": the magnetic configuration itself. As noted, the divertor will be subject to extremely intense fluxes of heat. Even if we can imagine managing the effects of heat on a machine like ITER, the availability of energy—or the lack thereof—remains a sticking point for future reactors, which will have to operate continuously. Situations will arise in which it is necessary to replace a damaged or aging divertor, since such equipment normally lasts for one year at a time. Paths now being explored involve optimizing the divertor's geometry and the technology of components, maximizing radiation in the plasma's peripheral regions, and incorporating artificial intelligence in order to manage and control the interface between the plasma and its material environment in real time.

Finally, if we follow the logic of previous chapters all the way through, the "second" form of magnetic configuration—the stellarator—offers clear advantages. As we saw in chapter 5, a stellarator doesn't need a plasma current, which means that no instabilities (either of the "disruption" type or edge localized modes) will affect performance. Since it doesn't have a plasma current, the stellarator doesn't need to recirculate energy, either. The operating point can lie at higher densities than is the case for a tokamak, and edge plasma conditions are a priori more favorable. A number of ITER partners, including Japan and Europe, are investing heavily in developing stellarator technology, which, to date, has been overshadowed by the tokamak. The German W7-X at Greifswald—a project supported at the transnational, European level that went into operation in 2016—is remarkable in this regard. The first wave of experiments has already shown results that live up to expectations and herald a particularly interesting phase of research. That said, the stellarator's technological complexity, in terms of both construction and maintenance, still must be evaluated before victory is declared or design plans are changed for the reactor of the future.

Even if a stellarator, and not a tokamak, ultimately forms the core of this device, all the advances from which ITER benefits over the next two decades will serve the same end goal; tokamaks and stellarators have a great number of features in common. The vast majority of fusion researchers are in agreement that ITER represents a necessary stepping-stone en route to the fusion reactor. The results obtained here will exert a key influence on the further course of their shared enterprise. Indeed, the project's name is quite fitting: in Latin, *iter* means "road."

On fusion road maps, DEMO is the usual designation for the stage that will follow ITER and integrate all the aforementioned elements into a more "industrial" approach. DEMO will also have the task of closing the thermal cycle and producing electricity from the heat collected. Doing so will make it possible to document more fully the competitiveness of fusion as an energy source of commercial interest. The most realistic timetables anticipate that such proof will be provided early in the last quarter of the twenty-first century. Additional research on materials and the overall efficiency of auxiliary heating systems—if results are favorable—will open the prospect of significant gains in terms of the costs of building and then operating the reactor commercially. Needless to say, all of this is predicated on the research community gradually handing things over to industry—a transition that will take some initiative.

In sum—that is, to condense an adventure that has unfolded over the course of a hundred years—we can classify the major stages of fusion (past, present, and future) in terms of so-called technical readiness levels (see figure 8.1).[1] In previous stages (epitomized by the

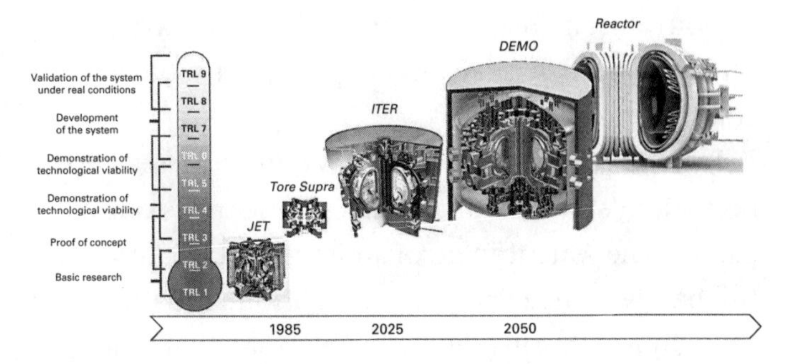

Figure 8.1
The major stages of fusion en route to the reactor. TRL, technical readiness level; JET, Joint European Torus.

JET and Tore Supra machines, which represented the height of endeavor for their day), researchers tested out the fundamentals of scientific feasibility and technological development. Now, ITER has the task of integrating these advances to demonstrate their scientific and technical viability as a whole on the new kind of nuclear installation where they will be employed. DEMO, the next stage, will address the industrialization of this process. Then, the first commercial reactor can go into operation (likely at the end of the century).

Other Uses of Fusion

As we reach the end of our journey, it is fitting—and important—to remark on the increased attention private investors are paying to a field that, until now, was not only a state matter but one reserved for a "club" of ultradeveloped nations in the position to undertake a colossal enterprise.

For the last ten years, magnetic fusion research has been flourishing at financially robust high-tech companies. The American aeronautics giant Lockheed Martin has launched its Compact Fusion initiative, and the Chinese fossil fuel giant ENN has its brand-new Fusion Research Institute. Research is also being conducted at start-ups financed by major venture capital companies—for instance, Tokamak Energy, in England. How should one interpret this development?

A few remarks are in order before turning to what can only be a personal evaluation of the situation. First, almost all of these initiatives are characterized by aggressive marketing campaigns for securing private capital, which tend to stress the high costs and extended lead times of so-called official approaches to magnetic fusion (that is, tokamak technology and the current stage of ITER operations). The argument of a more compact and cheaper solution is clearly meant to appeal to investors looking for medium-term gains in a dynamic market where money is not really a problem.

Thus, Lockheed Martin took the field in 2014 with the bold claim that it would be able to develop and market a minireactor that fits on the "trailer of a truck" within the next ten years or so. Other firms have announced that they're developing fusion reactors based on deuterium and helium-3—which offers the immense advantage of not requiring radioactive tritium and not generating neutrons (in other words, activation). We should note that such reactions require much higher temperatures than ones that employ a deuterium-tritium mixture, which means appreciably lower reaction efficiency. More cautious initiatives are focusing on alternative technical solutions for various subsystems, betting on a market that doesn't exist yet and a return on investments through patent licensing. The latter category includes efforts to develop tokamak magnets based on high-temperature superconductor materials, which would significantly lower the level of refrigeration necessary and open the possibility for operations with a higher magnetic field—two factors providing obvious benefits.

Engineers and researchers are revisiting the magnetic configuration, in particular. Their working assumption is that designs from the early days were abandoned too quickly and could be optimized much more effectively. Doing so involves small-scale experiments and following the same paths as taken by tokamak pioneers in the final decades of the twentieth century—retracing the steps that led first to JET and then to ITER. While the magnetic configuration is important, technological "building blocks" and their integration into highly complex systems prove just as vital. The course that such laboratories (and the industries behind them) are steering holds interest since they're counting on breakthrough and innovation, but it's also perilous. It's pretty well documented by now how long it takes to travel down this road.

Sixty-odd years of magnetic fusion research have witnessed a good number of technological spin-offs that industry can profitably employ, including applications for magnetized hot plasmas. To take just one example, some magnetic configurations might be used as very fast neutron sources that serve not to increase energy but rather to reprocess (and highly efficiently) some of the radioactive materials generated by fission plants; at present, many of these materials pose long-term storage problems. Such an application clearly merits attention from scientists and engineers working on fuel cycles.

After a fair amount of observation (and mistrust), it is now plain that the two worlds of fusion—state-sponsored and private—are drawing closer together; before long, collaboration to mutual advantage will no doubt take place. Only the state-sponsored approach guarantees that, if a new source of nuclear energy should emerge, it will conform to strict safety and security guidelines, and that these rules will be binding in the future. All the same, the somewhat sudden appearance of private investors and industry represents a sign—and a very positive one, at that—that an undertaking without parallel or precedent has passed a decisive milestone and a bright future lies in store.

* * *

In conclusion, fusion research for generating electricity has entered a new era. The extraordinary, worldwide consensus attending the collective enterprise of ITER symbolizes what is known as "scale-1 feasibility." The project has advanced thanks to a half-century's worth of coordinated research, men and women working toward the common goal of harnessing the energy of the stars to peaceful ends. ITER's success will not only crown the achievements of scientists and engineers. It will prove that it's possible to mobilize human beings from across the globe to face universal challenges. Let's hope that fusion will inspire further adventures and initiatives, including the preservation of the natural world and better distribution of wealth. Even though the scale of the enterprise, the complications that attend it, and inevitable setbacks may sometimes strike both researchers and the general public as reasons to give up, an undertaking like this surpasses and transcends the individual and individual efforts. Results to date confirm that remarkable progress has been made—more than enough reason to believe the "star-makers" will succeed.

Glossary

ASDEX-Upgrade Axially symmetric divertor experiment. Tokamak operated by the Max Planck Institute at Garching (near Munich) in Germany, in collaboration with the European fusion program; the main objective is to document scaling laws for confining a few dozen cubic meters of ITER-type tokamak plasma.

Auxiliary heating Any energy added to a fusion plasma by external means.

Bootstrap current Part of the plasma current generated by the plasma itself, tied in particular to its internal pressure profile. Under certain conditions, the bootstrap current will become dominant in a tokamak, opening the way for stationarity and significantly reducing the need for noninductive current generation.

Confinement The array of characteristics defining a plasma's performance as its capacity to trigger fusion reactions. May be understood in terms of both energy and particles.

Cyclotron heater Auxiliary heating based on injecting waves through antennas at the periphery of a fusion plasma; the frequency is adjusted to resonate with the rotational motion of particles around magnetic field lines. Depending on the target particles in question, scientists distinguish between electronic cyclotron heating and ion cyclotron heating.

DEMO In broad terms, a fusion machine that will follow ITER research; the main goal is to demonstrate the possibility of industrializing fusion reactors and producing energy. At several partner sites (including Europe and Japan), DEMO is currently at the preconceptual design stage, with no set timetable for construction.

Diagnosis Any method for measuring the physical quantities of a fusion plasma or its physical environment.

Disruption Very rapid and irreversible loss of magnetic configuration in a fusion plasma prompted by magnetohydrodynamic instability, which results in all magnetic and kinetic energy draining into housing structures.

Divertor The plasma-facing component most exposed to heat flux and particles; it is called by this name when the outer edge of the plasma's magnetic configuration presents a singularity known as the *x point* on the poloidal plane.

ELM, edge localized mode Magnetohydrodynamic instability at the outer edge of a fusion plasma in H-mode confinement, which causes a rapid drop of pressure profile; heat and particles then drain toward plasma-facing components.

Gyrotron Microwave tube for generating an electromagnetic wave injected into fusion plasma in order to heat it. Electromagnetic oscillation is generated by a beam of accelerated electrons passing through a cavity with corrugations at appropriate intervals. Gyrotrons are the main source of electron cyclotron heating in fusion, operating at a level between one hundred and two hundred gigahertz.

H-factor Amplification factor of a fusion plasma's energy confinement time, reflecting the transition between two states of confinement and generally observed above a certain level of overall power evacuated by the plasma. At low levels, the reference mode is called *L-mode*, with an H-factor set by convention at one. "Improved" confinement modes observed until now are *H-mode*, with an H-factor generally close to two, and/or advanced modes with H-factors between one and two.

H-mode Tokamak mode of operation with improved confinement (H-factor at about two), associated with the appearance of a pressure pedestal at the periphery of the plasma. First discovered on the ASDEX-Upgrade and subsequently observed on many other tokamaks.

The ITER Operating Point is dimensioned as an H-mode.

Hybrid heater (or lower hybrid frequency heating) Auxiliary heating based on injecting waves through antennas at the periphery of a fusion plasma; the frequency is adjusted to resonate with the motion of electrons along magnetic field lines. De facto, this method aims to generate current by noninductive means.

Ice cubes Deuterium, tritium, or helium cooled to form small solids of material for injection into a fusion machine in order to supply reagents.

IFMIF, International Fusion Materials Irradiation Facility Neutron irradiation facility for developing fusion materials. The operative principle is for accelerated deuterium nuclei to strike a liquid lithium target. The D-Li reaction produces neutrons with characteristics very close to those of D-T fusion. Currently in the proof-of-concept phase at the Rokkasho site in Japan.

ITER Tokamak under construction in Cadarache, France, under the supervision of the ITER International Organization; other partners are China, South Korea, the United States, Europe, India, Japan, and Russia. Its main objective is to provide scientific and technical proof of controlled deuterium-tritium fusion reactions by means of magnetism, with an energy amplification factor on the order of ten and reaction times on the order of one hour.

JET, Joint European Torus Tokamak operated by the European Union (Euratom) on British soil (Culham, Oxfordshire, UK) since the early 1980s. Its main objective is to define an operating point that can be extrapolated to the energy amplification projected for ITER, with plasma geometry very close to that of ITER. Although ten times smaller in volume than ITER, JET is the largest and most powerful tokamak in operation in the world; its outstanding feature is to allow tritium to be manipulated for experimentation.

JT-60SA Superconducting tokamak under construction at Naka (near Tokyo, Japan), with the goal of exploring so-called advanced plasma scenarios, which will provide future experimental campaigns with operating points that are more efficient and more stationary than those used to dimension ITER. JT-60SA is the successor to JT-60U and represents a joint effort between Japan and Europe.

JT-60U Japanese-operated tokamak located in Naka (near Tokyo) constructed for much the same purposes as JET. Shut down and dismantled in the 2000s to make way for JT-60SA.

Klystron Microwave tube for generating an electromagnetic wave for injection into a fusion plasma in order to heat it. The klystron produces electromagnetic oscillation from a beam of accelerated electrons passing through a cavity in the presence of a low-power wave with characteristics identical to the wave desired.

Klystrons are the main sources of hybrid heating in fusion, operating at levels around a few gigahertz.

Limiter The plasma-facing component most exposed to heat fluxes and particle fluxes. It is given this name in cases in which the outer edge of the plasma's magnetic configuration does not exhibit singularity on the poloidal plane.

Magnetic configuration Global magnetic field confining a fusion plasma generated by external components imposed by coils as well as internal components imposed by plasma current.

MHD, magnetohydrodynamics The branch of physics devoted to the stability of hot and magnetized plasmas on a large scale.

Neutral beam injection Auxiliary heating system for fusion plasma based on accelerating deuterium or tritium ions that are neutralized and then injected into the plasma. The energy of neutral atoms injected in this way is much higher than the average energy of ions in the plasma, thus causing the heating desired.

Noninductive current generation Any method for replacing all or part of the tokamak's transformer effect to maintain the plasma current. To varying degrees, all auxiliary heating methods have the capacity for noninductive current generation. Bootstrap current may also be described as noninductive.

Ohmic heating Intrinsic heating of plasma by an electric current flow through nonzero resistance associated with the Joule effect.

Plasma The fourth state of matter, typically obtained when heat surpasses tens of thousands of degrees and electrons are no longer bound to nuclei. The plasma state is present on a massive scale in the visible matter of the universe.

Plasma-facing component Any material element forming the wall of a fusion device subject to a direct flow of heat and particles from the plasma.

Poloidal Reference direction in a torus, tied to the magnetic field created by the plasma current. A poloidal plane is any vertical cross section of the torus.

Stellarator Magnetic configuration for fusion achieved entirely by means of external field coils.

Tetrode Microwave tube that amplifies low-level oscillation to generate an electromagnetic wave, which is then injected into a fusion plasma to heat it. In fusion, tetrodes represent the main source of ion cyclotron heating and operate in the vicinity of a few dozen megahertz.

TFTR, Tokamak Fusion Test Reactor Tokamak operated by the Princeton Plasma Physics Laboratory from 1982 to 1997. To date, TFTR and JET are the only tokamaks used for experiments with deuterium and tritium plasmas.

Tokamak Magnetic fusion configuration achieved by combining magnetic fields generated with vertical external field coils and a toroidal current flowing through the plasma.

Tore Supra Tokamak manufactured and operated by the CEA at Cadarache in collaboration with the European fusion program. Its principal objective is to develop technologies for operating fusion machines continuously (superconductive magnets, active cooling for plasma-facing components, auxiliary heating, diagnostic systems . . .). Plasma was first made with Tore Supra in April 1988.

Toroidal Reference direction in a torus along the plasma flow. The toroidal magnetic field is generated by the vertical field coils of the tokamak.

W7-X, Wendelstein 7-X Stellarator built and operated by the Max Planck Institute at Greifswald, Germany. W7-X is the largest superconducting stellarator in which all plasma-facing components will be actively cooled (eventually). Intended to document, in great detail, comparative stellarator and tokamak performance levels, possibly paving the way for a reactor with the same configuration. Plasma was first made with W7-X in December 2015.

WEST Since 2013, WEST has been the name for the Tore Supra tokamak, now reconfigured with a magnetic divertor and plasma-facing components in actively cooled tungsten walls. Plasma was first made with WEST in December 2016.

Notes

Chapter 1

1. See peakoilbarrel.com/world-energy-2017-2050-annual-report.

2. A radioactive element's *half-life*, or *period*, is the time it takes for the radioactivity of a given sample to decrease by 50 percent.

3. Nuclei with the same number of protons but a different number of neutrons are called *isotopes*. Two isotopes will have the same electrical charge, but different masses. Hydrogen has a single proton, deuterium has one proton and one neutron, and tritium one proton and two neutrons.

Chapter 2

1. Quoted in Sylvia Engdahl, ed., *The Atomic Bombings of Hiroshima and Nagasaki* (New York: Gale, 2011), 87.

Chapter 3

1. An *order of magnitude* stands for a factor of ten between two values of the same physical quantity. Thus, two orders of magnitude are a factor of one hundred, and so on. There are three orders of magnitude between one millimeter and one meter, and six between one millimeter and one kilometer.

2. The effective cross section is a physical magnitude directly connected to the probability of interaction between particles in the event of a reaction.

3. *Cold fusion* is the attempt to bring the nuclei of small atoms, such as deuterium and tritium, close enough to react without using a plasma—that is, under ambient conditions. Several approaches have been explored, from *muon-catalyzed* fusion in molecules whose electrons have been replaced by muons (which draw the nuclei much closer together) to efforts relying on the properties of crystal structures. Despite a few sensational headlines, no convincing demonstration has occurred to date.

4. *Sublimation* is a body's direct transition from the solid to the gaseous state.

Chapter 4

1. *Proceedings of the International Conference on the Peaceful Uses of Atomic Energy*, vol. 31 (UN: 1958), 27.

2. *Proceedings of the International Conference on the Peaceful Uses of Atomic Energy*, 20.

Chapter 6

1. The *tesla* (T) is the unit of magnetic induction in the International System of Units. For reference, the intensity of the earth's magnetic field lies at a few tenths of a millionth of one tesla.

Chapter 7

1. JT-60U, because it employed just deuterium, only reached breakeven virtually.

2. Among the many key players here, Bernard Bigot, whose involvement in ITER dates to 1996, warrants particular mention. At the time, Bigot was general director for research and technology at the Ministry for Education and Higher Learning; subsequently, he contributed

to the project as chief of staff for Claudie Haigneré and deputy cabinet director for Luc Ferry, then as high commissioner for atomic energy and general administrator of the CEA. On March 5, 2015, Bigot was elected to the post of ITER director general.

Chapter 8

1. This scale, initially proposed by NASA, seeks to quantify how close a given technology is to being used; it ranges from 1 (the idea) to 9 (the commercially available product).

Bibliography

Braams, C. M., and P. Stott. *Nuclear Fusion: Half a Century of Magnetic Confinement Fusion Research*. Bristol, UK: IOP, 2002.

Dinan, Richard. *The Fusion Age: Modern Nuclear Fusion Reactors*. Applied Fusion Systems Ltd, 2017.

Hawryluk, Richard, and Hartmut Zohm. "The Challenge and Promise of Studying Burning Plasmas." *Physics Today* 72, no. 12 (2019). https://physicstoday.scitation.org/doi/10.1063/PT.3.4363.

Herman, Robin. *Fusion, the Search for Endless Energy*. Cambridge: Cambridge University Press, 1990.

Mallonee, Laura. "Fusion Energy Gets Ready to Shine—Finally." *Wired*, April 21, 2020. https://www.wired.com/story/fusion-energy-iter-reactor-ready-to-shine.

McCracken, Garry, and Peter Stott. *Fusion: The Energy of Universe*. Amsterdam: Elsevier, 2005.

Morse, Edward. *Nuclear Fusion*. Cham, Switzerland: Springer, 2018.

Pacchioni, Giulia. "The Road to Fusion." *Nature*, September 24, 2019. https://www.scientificamerican.com/article/the-road-to-fusion.

Parisi, Jason, and Justin Ball. *The Future of Fusion Energy*. Hackensack, NJ: World Scientific, 2019.

Singer, Neal. *Wonders of Nuclear Fusion: Creating an Ultimate Energy Source*. Albuquerque: University of New Mexico Press, 2011.

Turrell, Arthur. *The Star Builders: Nuclear Fusion and the Race to Power the Planet*. New York: Scribner, 2020.

Index

Note: Page numbers in italic type indicate illustrations.